Yellowstone and Grand Teton National Parks and the Middle Rocky Mountains

Casper, Wyoming to Salt Lake City, Utah
July 20–30, 1989

Field Trip Guidebook T328

Leader:
J. David Love

Local Experts:
Gerald E. Nelson
Wayne Hamilton
James C. Coogan
Robert C. Palmquist
Roderick A. Hutchinson
Carl Ulrich
William G. Pierce
Robert B. Smith

American Geophysical Union, Washington, D.C.

Published 1989 by American Geophysical Union

2000 Florida Ave., N.W., Washington, D.C. 20009

ISBN: 0-87590-668-0

Printed in the United States of America

COVER ILLUSTRATION: Most of the gear in "Pioneer Geologists' Camp" was used for more than 40 years by David Love during work for the U.S. Geological Survey. Many of these items had been used previously by other U.S.G.S. field parties, some possibly as far back as the Hague Survey of Yellowstone National Park, 1886-1903. They had been stored in Miles City, Montana in 1904 and liberated for re-use in 1938. The packsaddle and U.S. Cavalry leather saddlebags were originally military equipment and the bags have waterproof canvas liners to hold wild game. The panniers were constructed of rawhide nailed to a hardwood frame. The gold pan belonged to David Love's father, who homesteaded in central Wyoming in 1896. The canteen is the type used in World War I. Other items include a Brunton compass, about 1915 vintage, used by Anaconda Mining Company geologists for underground mine mapping in the Butte, Montana, area. The pick and rock hammer were used by Laura W. McGrew, U.S.G.S. geologist, from 1952 to 1979. The artist, Phyllis Ranz, is Mrs. McGrew's niece and is a cartographer for the Wyoming Geological Survey.

Leader:

J. David Love
U.S. Geological Survey (Scientist Emeritus)
Box 3007, University Station
Laramie, WY 82071

Local Experts:

Gerald E. Nelson
Department of Geology
Casper College
125 College Drive
Casper, WY 82601

William G. Pierce
U.S. Geological Survey (Scientist Emeritus)
14380 Manuella Rd.
Los Altos, CA 94022

Roderick A. Hutchinson
Old Faithful
P.O. Box 2207
Yellowstone National Park, WY 82190

James C. Coogan
Department of Geology
University of Wyoming
Box 3006, University Station
Laramie, WY 82071

Robert C. Palmquist
P.O. Box 57
Shell, WY 82441

Wayne Hamilton
National Park Service
P.O. Box 213
Yellowstone National Park, WY 82190

Robert B. Smith
Department of Geology and Geophysics
University of Utah
Salt Lake City, UT 84112

Carl Ulrich
Fossil Station
Kemmerer, WY 83101

TABLE OF CONTENTS

IGC FIELD TRIP T-328
YELLOWSTONE AND GRAND TETON NATIONAL PARKS AND THE MIDDLE ROCKY MOUNTAINS

J. David Love
U.S. Geological Survey, Scientist Emeritus
Laramie, Wyoming

INTRODUCTION

The route of Field Trip T-328 through Wyoming is shown on the tectonic map (Figure 1), and an abbreviated schedule is presented in Table 1.

This trip is designed to show participants the granite-cored Laramide (Late Cretaceous-earliest Eocene) mountain ranges in the middle Rocky Mountains, and their various stages of burial by Cenozoic deposits and subsequent Quaternary exhumation. Mountain-flank structures involving Precambrian, Paleozoic, and Mesozoic rocks, the classic Heart Mountain detachment fault complex, and the rootless overthrust mountain ranges of the Wyoming-Utah-Idaho thrust belt are traversed.

The Gas Hills, one of the largest "boom and bust" uranium open-pit mine areas in the United States is examined. Oil and gas fields are seen in two of the major intermontane basins (Wind River and Bighorn), and also in the overthrust belt that, in the last 15 years, has been found to be fabulously productive. The Absaroka volcanic field (Eocene) and the Late Cenozoic igneous, thermal, and seismic feature of Yellowstone National Park are traversed, as well as the related Hebgen earthquake and landslide phenomena created in 1959.

The trip includes the Tetons, the youngest and one of the most spectacular fault-block ranges in the Rocky Mountains, still rising, and superimposed across an ancestral Laramide compressional range. Jackson Hole to the east has been sinking during the last 20 million years and in it is trapped one of the thickest and most complete Miocene and younger nonmarine sedimentary sequences in North America. Quaternary glacial and lacustrine deposits, superimposed rivers, Holocene faults, and seleniferous rocks are featured.

Stops also include a Green River (Eocene) fossil fish quarry and an Eocene tropical fossil leaf locality where participants can collect their own specimens. Another stop is near the famous Shell dinosaur quarry in Upper Jurassic rocks.

The geology along the route of Field Trip T-328 is shown on segments of the geologic map of Wyoming (Love and Christiansen, 1985). Detailed articles on the local geology, economic resources, and human history are contained in 39 Annual Field Conference volumes of the Wyoming Geological Association, and references contained therein. In addition, two new books present the general geology of Wyoming (Blackstone, 1988; Lageson and Spearing, 1988). Brief descriptions of the composition, thickness, geologic history, and economic resources of rocks in and marginal to the Wind River and Bighorn Basins (Love, 1988) and in Jackson Hole (Love, 1977) are likewise available. Pertinent publications on the thrust belt area of western Wyoming and northeastern Utah are listed in the narrative descriptions.

TABLE 1 Summary of Daily Events of IGC Trip T-328

DAY	LOCAL EXPERT & AFFILIATION	STOP	GEOLOGY	EVENING: LODGING	EVENING: PROGRAM
0	Dr. G.E. Nelson Casper College and W.G.A. officials	–	Arrival during day (July 20). Ice-breaker hosted by W.G.A., and talk by Dr. Nelson in evening	Casper Hilton	Geologic setting of Casper area
1 21st	J.D. Love, U.S.G.S. Scientist Emeritus	Alcova overlook	Section of Precambrian to Mesozoic and Tertiary rocks, NE flank of Granite Mountains		
		Split Rock overlook	Precambrian core of Granite Mountains partly buried by Split Rock Formation (Miocene)		
	Pathfinder Mines staff	Gas Hills uranium area	Lucky Mc uranium pit in Wind River Formation (Eocene) and general geology of area		
	J.D. Love	Boysen flora	Tropical swamp flora of middle Eocene age (Wagon Bed Fm.). Allow time for collecting		
		Wind River Canyon	Precambrian and Paleozoic stratigraphy and faulting in 3,000′ canyon cut by Wind-Bighorn River superimposed across Owl Creek Mts.		

		Overnight		Holiday Inn Thermopolis for some, adjacent AAA for others	Optional swim or tour of State Park
2 22nd	J.D. Love and Dr. R.C. Palmquist, N.W. Community College	Shell Canyon	Precambrian, Paleozoic, Mesozoic, and Quaternary stratigraphy, structure, and glacial history. Dinosaur quarry area noted		
		West Burgess Jct. overlook	Precambrian and Oligocene stratigraphy, regional structure, and overview of Bighorn Basin geomorphology		
		Five Springs	Stratigraphy and structure of steeply folded and faulted ramp along west flank of Bighorn Mts.		
		Optional Bighorn Canyon Rec. area	Superposition of Bighorn River across northern Bighorn Mts. and display of Paleozoic rocks		
		Overnight		Buffalo Bill Village, Cody	Buffalo Bill Museum, optional
3 23rd	Dr. W.G. Pierce & J.D. Love, U.S. G.S.	Rattlesnake Mountain	Thrust-faulted anticline with Precambrian core flanked by Paleozoic and Mesozoic rocks, river terraces and thermal features		
		Dead Indian Hill	Close-up view of Pal. and Mes. stratigraphy and relation of thrust masses to Pat O'Hara anticline and Heart Mountain klippe		
		Heart Mountain	View northeast of Heart Mountain detachment mass and stratigraphy of both Pal, Mes., and Tertiary rocks and displacement, forces of emplacement, interpretations		
		Clark Fork Canyon	Stratigraphy and structure of east margin of Beartooth Mountains, lower Eocene conglomerates derived from it, and glacial events		
		Overnight		Red Lodge, MT	Optional
4 24th	Same leaders as for day 3	Beartooth Plateau Overlook	Precambrian geology, Beartooth physiography, glacian stages		
		Hunter Peak	Close view of detachment masses of Paleozoic rocks on Heart Mountain thrust plate		
		Cathedral Cliffs	View of type Cathedral Cliffs conglomerate and its relation to Heart Mountain thrust emplacement		
		Cooke City (optional)	Old gold, silver, and copper mining town. Most mineralization in Cambrian strata intruded by Tertiary igneous rocks		
		Soda Butte (optional)	First evidence of modern thermal activity, Eocene volcaniclastic stratigraphy, fossil forests		
		Overnight		Gardiner, MT	Free
5 25th	Wayne Hamilton, Geol., Yellowstone Park and J.D. Love	Mammoth Hot Springs	Quaternary travertine, thermal features, radon, Mt. Everts stratigraphy, Bunsen Peak		
	Rick Hutchinson, Geol., Yellowstone Park	Norris Geyser Basin	Thermal features, both pre- and post-Hebgen earthquake		
	J.D. Love	Hebgen Earthquake area	Slide, dam, fault scarps, lake, and other features related to 1959 earthquake		
	Rick Hutchinson	Old Faithful	Thermal features, both pre- and post-Hebgen earthquake		
		Overnight		Canyon Village	R B. Smith (possible) for seismic talk

6 26th	Wayne Hamilton and J.D. Love	Yellowstone Falls and Canyon	Scenery views of falls and canyon and igneous and glacial geologic history of canyon area		
	Wayne Hamilton	Yellowstone Lake	Thermal features, lake shore deposits, tilting of lake basin, glacial history		
		Overnight		Wort Hotel, Jackson	R B. Smith, or free
7 27th	J.D. Love	Jenny Lake	Photo stops and geologic history of Tetons and Jackson Hole valley floor. Stop at Teton Park Hqs. for publications		
		Teton Village tram	Afternoon trip up tram for geologic history of region and physiographic overviews		
		Overnight		Same as for day 6	Optional
8 28th	J.D. Love	Gros Ventre Slide and valley	Geology of Gros Ventre Slide and flood, and valley upstream from slide		
		Teton Pass	Contrasting stratigraphy of thrust belt and foreland		
		Granite Hot Springs (optional)	Traverse through leading edge of thrust belt, northern Green River Basin overridden by both foreland and thrust belt		
		Overnight		Same as for day 7	Optional
9 29th	Jim Coogan, U. of Wyo., and J.D. Love	Astoria Hot Springs	Darby thrust fault and thrust belt stratigraphy and thermal features		
		Little Greys River anticline	Mesozoic stratigraphy, Little Greys River anticline and view of Absaroka thrust plate front		
		Alpine	General view of thrust belt structure modified by Late Cenozoic normal faulting		
		Salt Flat	Core of Afton anticline, salt springs, and decollement in Preuss Formation; salt flowage problems		
		Raymond Canyon	Structure of Crawford ramp and Paleozoic and Mesozoic stratigraphy and structure; vanadium experimental mines during WW II		
	Carl Ulrich quarry owner and Jim Coogan	Fossil Fish quarry	Stratigraphy and fossils of Green River Formation. Allow time for collecting fossils. Setting of Fossil Basin during thrusting (see Lamerson cross section)		
		Overnight		Kemmerer, WY	Possibly talk on thrust belt by Coogan
10 30th	Jim Coogan and J.D. Love	Echo Canyon	Thousands of feet of coarse conglomerate unconformably overlain by Wasatch Formation		
		Devils Slide	Evanston Formation (Cret-Pal) overlapping Paleozoic and Mesozoic rocks (see cross section)		
	Coogan, and R.B. Smith?	Weber Canyon	Archean core of Wasatch Range and its relation to previously-observed thin-skinned thrusts		
	R.B. Smith or a designate, U. of Utah	Wasatch Mountain Front	Anatomy of mountain front faults, Lake Bonneville history, seismic and real estate problems in densely populated area		
		Salt Lake City	Disband		

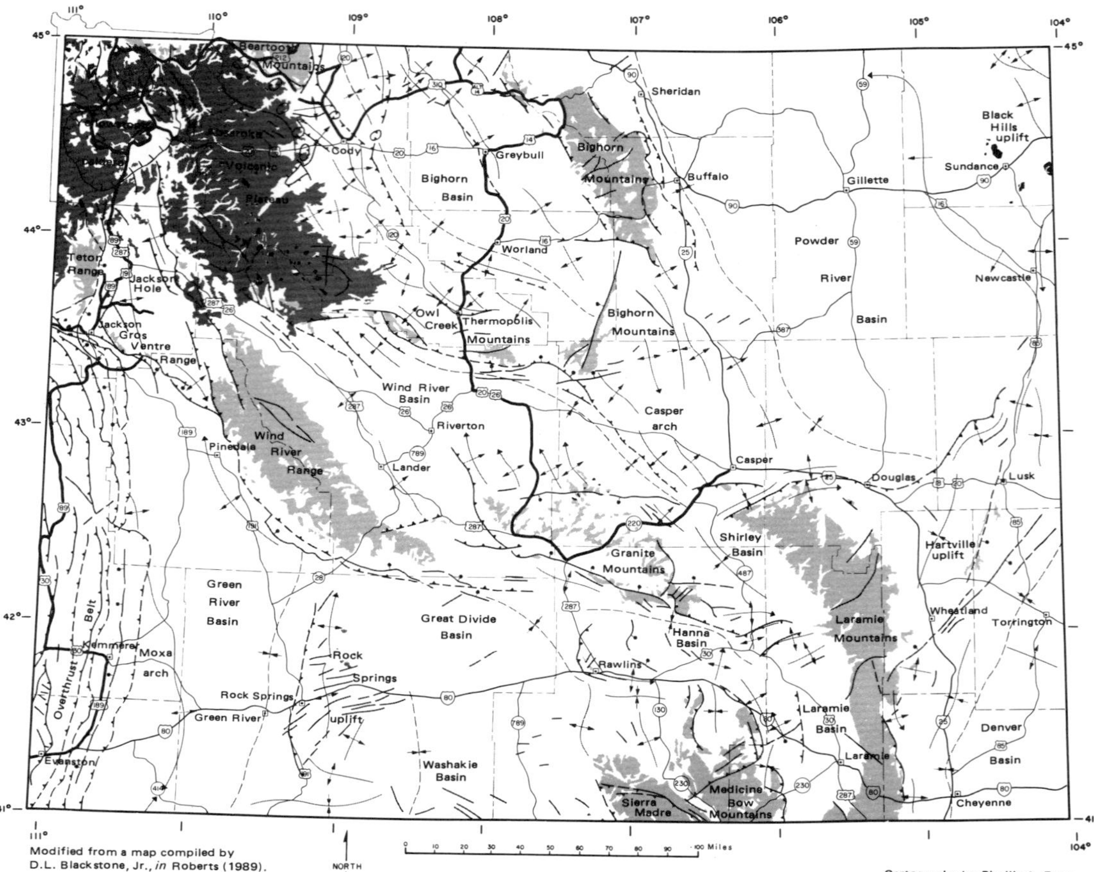

FIGURE 1 Tectonic map of Wyoming showing the route of Field Trip T-328 (heavy black line) across mountains, basins, volcanic areas, and the Wyoming overthrust belt (modified from Blackstone, 1988). Types of faults, anticlines, and synclines are shown with conventional symbols. Precambrian rocks have light gray and volcanic rocks have dark gray patterns.

DAY 1

The city of Casper (the largest city in Wyoming, population approximately 51,000) is located on a broad area of gently-dipping gray soft Cody Shale (Upper Cretaceous) overlapped by Pleistocene-Holocene dune sand (Figure 2). The headquarters for this field trip are adjacent to an interesting and significant prehistoric Indian bison kill site (Frison, 1978), about 10,000 years old. The bison were apparently driven into the soft sand dunes where they floundered and were easily butchered. Figure 3 shows the abundance of bones, some articulated, others scattered during the butchering process.

On the skyline south of Casper is Casper Mountain, rising 2,800 feet above the town. As is shown on Figure 4, a reverse fault of Laramide age (Late Cretaceous-earliest Eocene) has offset the Precambrian, Paleozoic, and Mesozoic rocks about 5,000 feet.

From Casper, the field trip route goes southwest for about 35 miles to Stop 1. The traverse crosses, nearly at right angles, Cretaceous, Jurassic, Triassic, Permian, and Pennsylvanian rocks folded in a series of northwest-trending anticlines and synclines (Figure 2). The first part of the route, along the North Platte River, coincides with the Oregon Trail (1840s to 1860s). During this interval, as many as 50,000 people per year migrated westward by wagons, horseback, and on foot, seeking land and fortune along the Pacific coast. This is also the route of the Pony Express (1860-61) that carried mail horseback across the United States before the transcontinental railroads were completed.

Figure 5 is a detailed map of the area near Alcova Reservoir (a U.S. Bureau of Reclamation irrigation storage reservoir), and can be used to identify the various stratigraphic formations. For convenience in orientation, the thick red sequence is the Chugwater Formation of Triassic age.

Pioneer geologist F.V. Hayden and his field party, while surveying the Wyoming Territory, camped at Red Buttes beside the field trip route along the North Platte

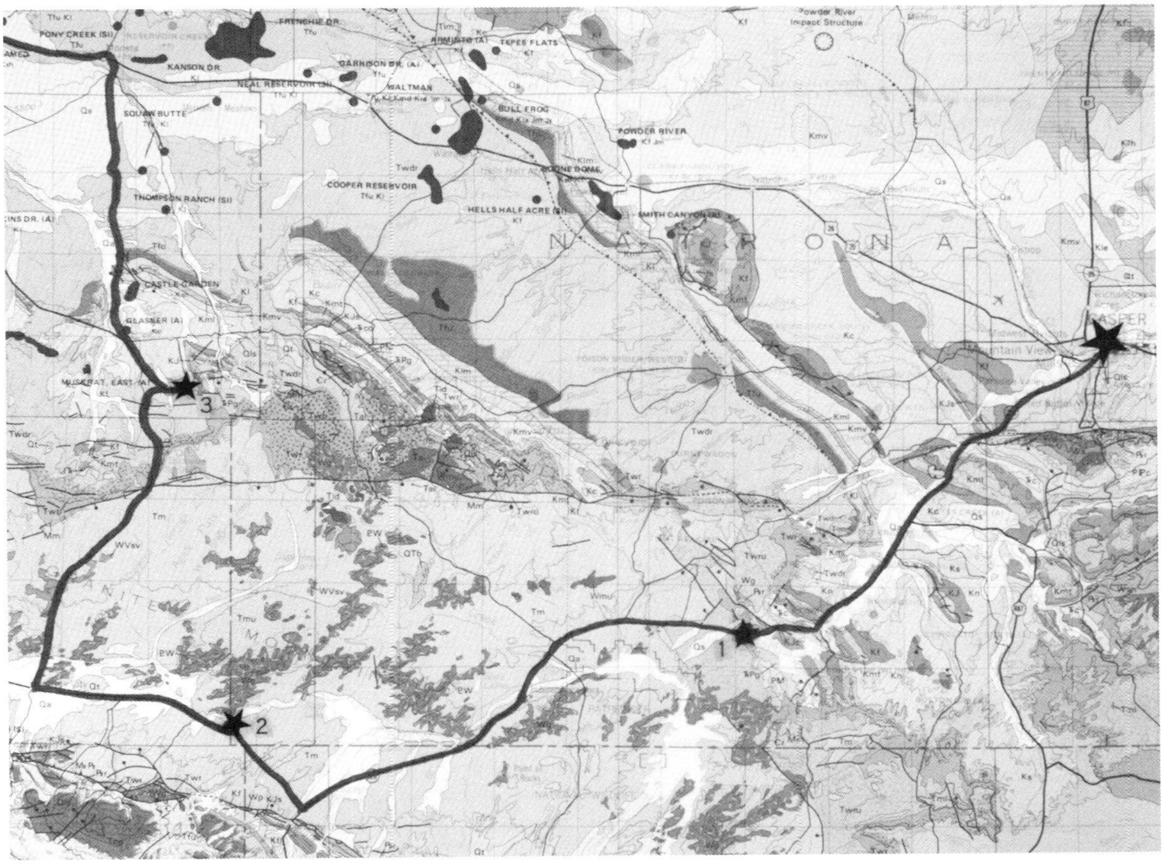

FIGURE 2 Segment of the geologic map of Wyoming (Love and Christiansen, 1985) showing the route of the field trip (heavy black line) on part of Day 1. Large star indicates overnight stop at Casper. Smaller numbered stars show stops along route. For geologic symbols and line patterns see source map. Squares are approximately six miles on a side. Black named areas are oil and gas fields.

River on Aug. 24, 1870 (Figure 6). He wrote a vivid and remarkably accurate description of the geology of this area, the first scientist to do so (Hayden, 1872, p. 27-28).

Stop 1

Stop 1 on the plateau margin west of the Alcova Reservoir (Figure 7) provides a view of the Paleozoic and Mesozoic stratigraphy and structure that have just been traversed. These strata were folded, and in some instances faulted, during the Laramide Orogeny (Late Cretaceous-earliest Eocene). The nearly flat-lying Oligocene and Miocene strata on the plateau to the west can be seen from this stop. These rocks were deposited across mountainous topography developed on the upturned edges of the older strata. The Oligocene rocks (White River Formation) consist of white blocky tuffaceous claystone. This formation is overlain by an arkosic conglomerate derived from the adjacent Granite Mountains and may be of either Oligocene or Miocene age. Overlying it is the Split Rock Formation, a thick sequence of gray sandstone, tuff, and thin limestones of Miocene age (Love, 1970).

For the next 45 miles to Stop 2, the route is along the partly-buried Granite Mountains. A series of "soda lakes" can be seen on both sides of the highway. Salts, chiefly Na_2SO_4, Na_2CO_3, and lesser amounts of NaCl, were leached out of the Split Rock Formation and deposited in undrained wind-blown depressions. The salts were mined from 75 to more than 100 years ago but there has been no exploitation in recent years.

The historic dome-shaped granite knob called Independence Rock can be seen on the left side of the highway directly south of the highway bridge across the Sweetwater River. This was the "pioneer register" along the Oregon Trail where emigrants carved their names on the 2,600-million-year-old granite, chiefly between the 1830s and 1870s. This and the Devils Gate, a spectacular sheer canyon cut by the Sweetwater River through another granite knob six miles southwest of Independence Rock, are some of the exposures of the partly-exhumed crest of the Granite Mountains.

One reason for routing Field Trip T-328 across the Granite Mountains is because they are unique in Wyoming. They remain partly buried by upper Cenozoic sedimentary deposits, whereas other mountain ranges

FIGURE 3 Bison kill site 10,000 years old in sand dune at Casper. Photography by George Frison.

that will be observed later in this trip have been almost entirely exhumed and their Cenozoic sedimentary record thus destroyed.

The following discussion is abstracted from Love (1970, p. C-1–C-3). Preservation of the sedimentary strata across the Granite Mountains was made possible by collapse of almost the entire uplift during late Cenozoic time, either prior to or contemporaneously with the epeirogenic uplift that in other parts of Wyoming, launched the present cycle of regional degradation.

Stop 2

The Precambrian rocks in the area of the field trip near Stop 2 consist mainly of pink and gray granite cut by abundant mafic dikes (Figure 8). Tertiary strata surround and overlap the Granite Mountains on all sides. The Precambrian core was exposed during Paleocene time but furnished much less arkosic debris than it did in the Eocene. The most violent movements of the Laramide Orogeny in central Wyoming came during earliest Eocene time. The Granite Mountains rose as a block 90 miles long and 30 miles wide. On the north side, along the field trip traverse, lower Eocene strata are represented by the Wind River Formation, derived by erosion of the adjacent uplifted mountain core. These strata are present along the entire north flank of the mountains. The Wind River consists of as much as 5,000 feet of conglomerate, arkosic sandstone, and variegated mudstone. It is the host for large uranium deposits in the Gas Hills area.

The mountainous topography developed during early Eocene time was progressively buried with very little modification from middle Eocene through Miocene time by debris that was not locally derived, but which came from distant volcanic vents (except for the relatively small ones of middle and late Eocene age in the Rattlesnake Hills 20 miles northeast of Stop 2). Thus, the highest granite knobs along the route (Figures 7 and 8) show essentially the same topography that one could have seen 50 million years ago!

The east-trending Split Rock syncline developed during

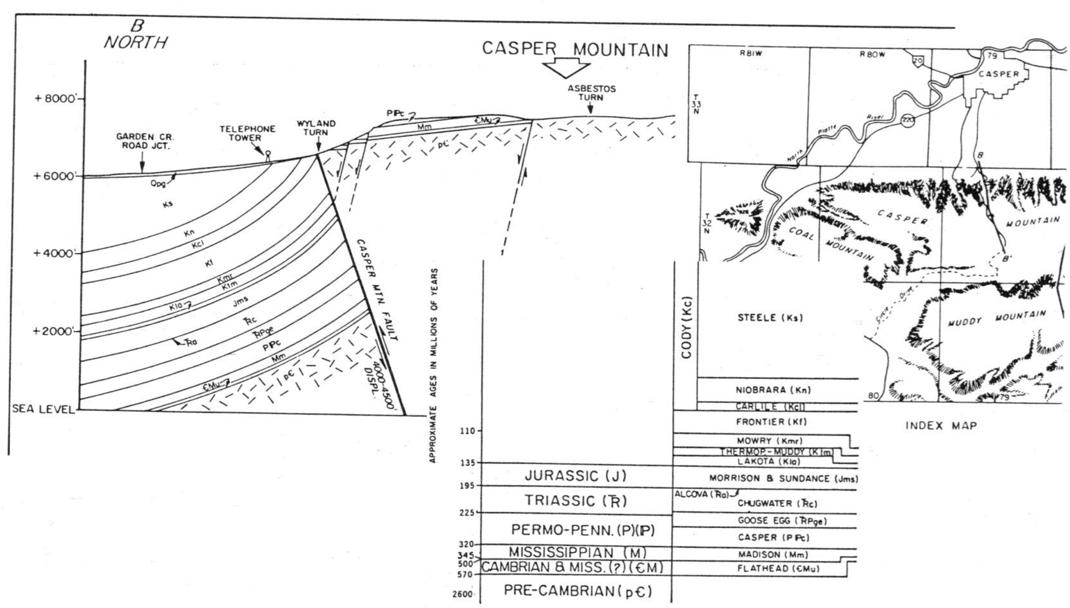

FIGURE 4 Cross section showing structure of Casper Mountain. Qpg is pediment gravel. Adapted from Jenkins and Rea *in* Van Burgh (1978).

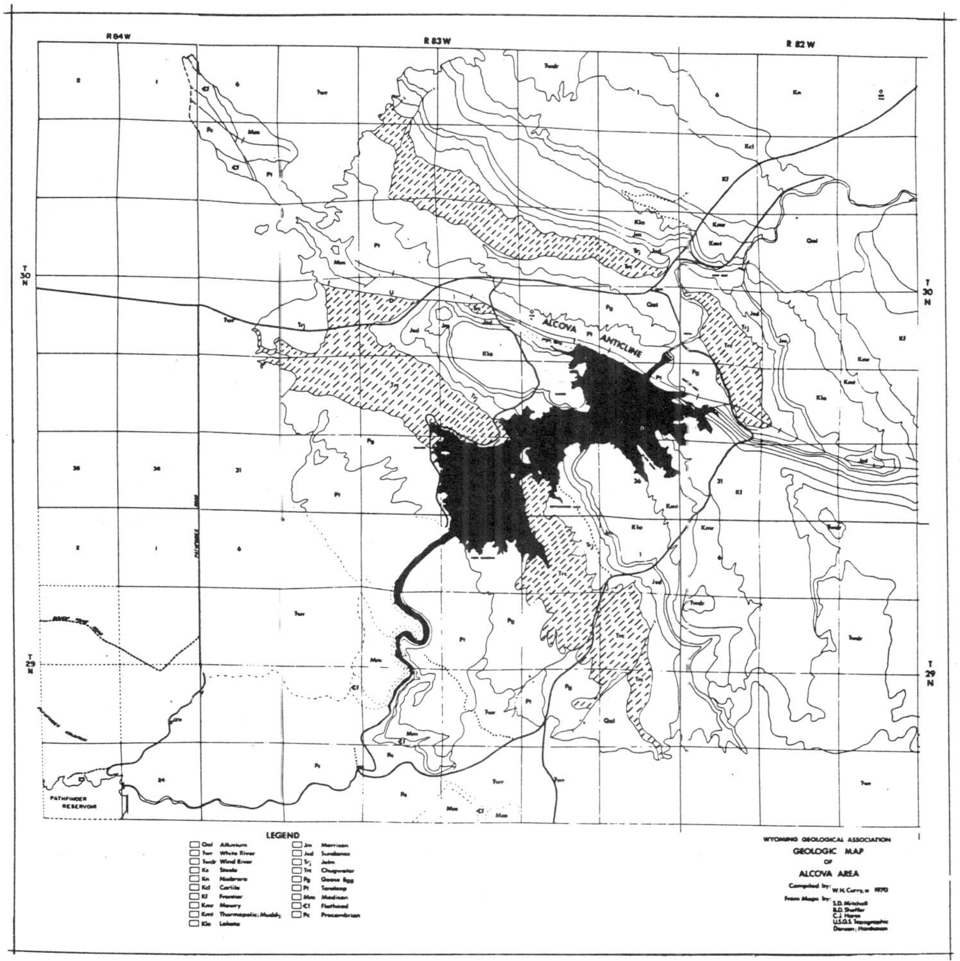

FIGURE 5 Geologic map of Alcova area on the way from Casper to Stop 1. Geology from W.H. Curry III *in* Knittel (1974).

Miocene time along what had formerly been the crest of the Granite Mountains. In it was deposited the Split Rock Formation. This formation is a vast accumulation of wind-blown and waterlaid sand nearly 3,000 feet thick. It buried all but the highest parts of the peaks. The thickest strata are along the route between Stop 2 and Jeffrey City. In the deepest part of the syncline, the base of the Split Rock Formation is only 3,000 feet above sea level (Love, 1970, Pl. 6), and is still sinking.

In late Miocene time, Moonstone Lake formed in the Split Rock syncline. More than 1,000 feet of tuffaceous strata, making up the Moonstone Formation, accumulated in and adjacent to this lake (Figure 9). The presence of diatoms of saline habitat, salt casts, and salt concentrations indicate that the lake had little or no external drainage. Some lacustrine strata in this sequence are unusually rich in uranium and thorium and are believed to be source rocks for part of the uranium present in the Gas Hills uranium district (Stop 3; Figure 10).

Elsewhere in Wyoming, regional uplift, probably during late Miocene to Pleistocene time, started the present cycle of degradation in most of central Wyoming. The Granite Mountains area, however, is unique because reactivation of movement along the North Granite Mountains fault system (Figure 2) and more widespread movement along the South Granite Mountains fault system (six miles south of Stop 2, Figure 2), caused the mountain block to drop farther. The regional effect of a diminished gradient of the Sweetwater River, the only major stream along the Granite Mountains, resulted in minimal erosion of the soft sediments surrounding the granite knobs. Therefore, only about 1,000 feet of the buried mountain crest was exhumed. This is the history of the present landscape shown in Figure 9.

From Stop 2, the route continues west-northwest on the Split Rock Formation to Jeffrey City and then turns north onto a major graveled (but not paved) uranium all-weather haulage road to the Gas Hills uranium district (Stop 3). About 20 miles north-northeast of Jeffrey City, the road crosses the North Granite Mountains fault system (Figure 2) and descends into the Wind River

Stop 3

Stop 3 is at the Lucky Mc uranium open pit mine (Figure 10). The proven uranium area in the Gas Hills

FIGURE 6 Hayden Geological Survey party at Red Buttes along the North Platte River, August 24, 1870. F.V. Hayden is seated at table, facing camera. Photographer W.H. Jackson is standing at right. U.S. Geological Survey photograph by W.H. Jackson.

district is about 15 miles long (northeast to southwest), four miles wide, and includes about 60 square miles. Three separate parts of it were discovered independently within a two-week period in September, 1953, by Neil McNeice, a prospector (after whom the Lucky Mc mine was named), by P.T. Jenkins, a uranium geologist, using an airborne scintillator, and by J.D. Love for the U.S. Geological Survey (Love, 1954). The uranium is in the Wind River Formation (early Eocene). Until this discovery, the conventional wisdom in the geologic profession was that most major uranium deposits were related to hydrothermal activity. These were not. The uranium probably was in part leached from tuff in the Moonstone Formation and in part from arkoses in the Wind River Formation derived from Precambrian granite. During the following 30 years (until the Three Mile Island accident curtailed the uranium industry), approximately 25 million tons of ore yielding 8.5 million pounds of uranium oxide were produced (Love, 1988a, p. 200). To the east of the Lucky Mc mine, a steeply folded northwest-plunging anticline in Triassic, Jurassic, and Cretaceous rocks emerges from under the cover of horizontal Tertiary strata. Table 2 summarizes the stratigraphic section in the Wind River Basin. The geologic history of the basin is summarized elsewhere (Love, 1988a, p. 199-200). From the mine, the route follows a paved road northwestward for about six miles and then turns north on an unimproved but graded road for about 20 miles to Moneta where it is again on pavement. (In the event of rain, the road to Moneta is not good and the route will be diverted northwestward to Riverton and then eastward back to Shoshoni (Figure 11), entirely on pavement.)

About six miles north on the Moneta road, the route crosses the cattle ranch where your leader grew up. East of this site, the conspicuous gray sandstone cliffs are in the Paleocene Fort Union Formation. Four miles farther north, on the west side of the road is a small area of red and white candy-striped unvegetated badlands in the Wind River Formation. This is a different facies from the basin-margin arkoses seen at the Lucky Mc mine in the same formation. The badlands contain small vertebrate fossils of early Eocene age. Beginning about six miles south of Moneta and extending north for four miles, the route crosses one of the longest (about 115 miles) stabilized sand dune areas in the United States (for the entire extent see Love and Christiansen, 1985). It is vegetated and inactive and its age is not known but, based on its similarity to the Casper dunes, it is probably at least 10,000 years old and probably older.

About ten miles north of Moneta is the Madden-Lost Cabin gas field with a producing area of more than 50 square miles. Productive horizons of low-sulfur gas are in lower Eocene, Paleocene, and Upper Cretaceous sandstones and large quantities of high sulfur gas occur at a

FIGURE 7 Angular unconformity between flat-lying Oligocene and Miocene rocks on the Sweetwater Plateau and northeastward-tilting Mesozic rocks in the northern part of the area shown in Figure 5. Photograph by John S Shelton.

depth of about 24,000 feet in the Madison Limestone (Mississippian). The Wind River Basin has a long history of oil and gas production, beginning as far back as 1883 when the first oil well was drilled in the Dallas field, 45 miles southwest of Shoshoni. Since then 113 named and productive fields have been discovered in and adjacent to the basin. Cumulative production is estimated to be more than 377,400,000 barrels of oil and about 1.2 trillion cubic feet of gas (Barlow and Haun, 1978; updated through 1983; Love, 1988a). Coal resources are estimated at approximately one billion tons. Cumulative coal production through 1973 (the last year of recorded production) was about 4 million tons (Glass and Roberts, 1978).

Between Moneta and Shoshoni the ledge-forming tan sandstones on the north side of the highway are in the upper part of the Wind River Formation, which here is more than 4,000 feet thick. From Shoshoni the route turns north toward the Owl Creek Mountains (Figure 11). On the west side of the highway is the Boysen Reservoir, a Bureau of Reclamation water storage project that dammed the Wind River, the major stream draining the Wind River Basin. In this area, the depth to Precambrian rocks is about 25,000 feet, yet in the Wind River Canyon straight ahead at eye level, they are exposed.

Stop 4

About 10 miles north of Shoshoni is Stop 4. Leave the bus on the edge of the highway and walk horizontally about 1,500 feet north to an abandoned highway road cut in green and gray tuffaceous claystone and sandstone of the middle Eocene Wagon Bed Formation. Thin carbonaceous shale and coaly beds here contain abundant and beautifully preserved fossil leaves, carbonized conifer wood fragments, and sparse seeds and insect wings. This is the site of the Boysen flora which is widespread in middle Eocene rocks in other parts of Wyoming. This locality was first described by Love (1939, p. 71), later by R.W. Brown (written commun. to J.D. Love, March 15, 1945), and by H.D. MacGinitie (written commun. to J.D. Love, Jan. 31, 1964; published by MacGinitie in 1969, p. 66). It is a tropical flora consisting chiefly of palm, *Lygodium,* and soapberry, but with lesser numbers of many other species, plus abundant pollen. This site provides an opportunity to collect leaf specimens of museum quality. Figure 12 shows a generalized cross section (Blackstone, 1988) from the Owl Creek Mountains to the Wind River Basin in this area. The gas well is located 2,700 feet northeast of Stop 4.

FIGURE 8 Split Rock, a pioneer landmark on the Oregon Trail near Stop 2 is composed of granite about 2,700 million years old, intruded along a fracture by a softer, more easily eroded black mafic dike, likewise of Precambrian age. The Precambrian knobs of the Granite Mountains were buried by the Split Rock Formation (Miocene) and then exhumed during Quaternary time. The configuration of the crest of the mountains has been essentially unmodified since Miocene time. View is west across the Sweetwater River. Photograph by J D. Love, Sept. 23, 1958.

FIGURE 9 Oblique air photograph showing type section of the Moonstone Formation (upper Miocene) deposited on a surface of high relief developed on the Precambrian core of the Granite Mountains prior to and during late Miocene time. View is north. Indicated are Moonstone Formation (A), with position of middle part of type section on White Ridge (B), Lone Mountain at upper right corner (C), Split Rock Formation (D). Photograph by P.T. Jenkins and L.P. House.

FIGURE 10 Air oblique view northeast near Stop 3 showing magnitude of uranium strip mines in the nearly horizontal Wind River Formation (lower Eocene) only five years after the discovery of the area. Indicated are Jurassic rocks dipping 20° west of the west flank of the Dutton Basin anticline (A), Lucky Mc camp (B), Vitro camp (C), Vitro strip mine (D), Lucky Mc strip mine at Stop 3, showing dark uraninite ore zone at bottom of pit (E), Fraser-Lemac strip mine (F). Mines are in sec. 26, T.33N., R.90W. Photograph by P.T. Jenkins and L.P. House, showing extent of stripping as of August 13, 1959.

Stop 5

Stop 5 is three miles farther northwest at the entrance to the Boysen Camp near the south end of Wind River Canyon. Figure 13 is an air oblique view of the canyon. The highway coming into view at the south boundary of the photograph is just north of Stop 5. Between Stops 4 and 5, the Wagon Bed and Wind River Formations lap northward onto the steeply folded and faulted Paleozoic and Mesozoic rocks along the south margin of the Owl Creek Mountains. Masses of highly brecciated Paleozoic rocks have been thrust southward onto Mesozoic strata and both sequences have been cut by younger normal faults. Details of the stratigraphy and structure here are shown on maps by Fanshawe (1939) and Tourtelot and Thompson (1948).

The jumbled terrain in the lower half of Figure 13 is composed of folded and faulted Paleozoic and Mesozoic strata. The dark Precambrian rocks in the center are on the north side of the Boysen fault (Figure 12). The route traverses the Precambrian rocks at the bottom of the canyon north of the fault for more than two miles and crosses magnificent exposures of the intricately tangled igneous and metamorphic rocks. These are chiefly amphibolite and felsic gneiss of volcanic origin, quartz pegmatites, and granite. Most are probably older than 2,700 million years.

The canyon walls on both sides of the river (whose name changes from Wind River to Bighorn River in the middle of the canyon – hence the designation "Wedding of the Waters") display details of Cambrian, Ordovician, possibly Devonian, thick Mississippian, Pennsylvanian, and Permian sequences. The canyon is nearly 3,000 feet deep (7,556 to 4,596 feet) at its deepest part. So – why does the river cut a canyon across the Owl Creek Mountains from the much lower Wind River Basin on the south and the still lower Bighorn Basin on the north? The river was superimposed from a high-level surface of Cenozoic strata that filled both basins and buried the Owl Creek Mountains (for this history, see Love, 1988b, p. 200-204). The time of canyon-cutting probably began in the late Cenozoic.

TABLE 2 Paleozoic, Mesozoic, and Teritary rocks within and along the margins of the Wind River Basin. (For formation names, correlations, and areas of outcrops, refer to Love and Christiansen, 1980; 1985; after Love, 1988.)

Geologic Age	Thickness ft (m)	Description
Miocene	0-900 (0-274)	White tuffaceous sandstone, thin white limestone, vitric tuff, gray claystone; conglomerates at top and base.
Oligocene	0-600 (0-183)	Claystone and siltstone, white to pale pink, massive, blocky, tuffaceous. Thin beds of vitric tuff; locally derived conglomerates near top and base.
Late and Middle Eocene	0-2,500 (0-762)	Along east and south margins: claystone, siltstone, and sandstone, green, yellow, and gray, tuffaceous. Along northwest margin: upper part is olive drab to green hard slabby tuff; lower part is highly variegated soft bentonitic tuff and claystone. Some strata are highly seleniferous and gray to brown.
Early Eocene	0-9,000 (0-2,743)	Variegated claystone, siltstone, sandstone, and conglomerate; very coarse conglomerate locally at base and top. These strata are the chief host for uranium in the southeastern part of the basin. In the deepest part of the basin there is 2,500 ft (762 m) of black and gray shale near base. In the northwestern part there are large detachment masses of Paleozoic and Mesozoic rocks near base.
Paleocene	0-8,000 (0-2,438)	Gray and brown, generally noncalcareous claystone, shale, and sandstone with some lenticular conglomerates and a few coal beds. In the northeastern part of the basin, the upper section contains a brown and gray shale sequence 2,500 ft (762 m) thick, possibly marine.
Late Cretaceous	5,600-8,000 (1,707-2,438)	From top down: gray and buff sandstone, gray to black shale, and thin coal beds (thickness 0-6,000 ft; 0-1,829 m); white to brown soft tuffaceous sandstone, thin carbonaceous shale and yellow bentonite, grading eastward into marine shale (thickness 800-1,300 ft; 244-396 m); white to gray sandstone, carbonaceous shale, and coal (thickness 550-2,000 ft; 168-610 m); gray marine shale with thin sandstones (thickness 3,000-4,700 ft; 914-1,433 m); gray marine sandstones and brown to black shale (thickness 500-1,000 ft; 152-305 m).
Early Cretaceous	525-1,250 (160-381)	Marine and brackish-water siliceous shale, bentonite beds, and tan sandstone; nonmarine sandstone and variegated claystone near base; basal cliff-forming conglomerate.
Jurassic	300-850 (91-259)	Uppermost 100-200 ft (30-61 m): nonmarine gray sandstone and variegated claystone; middle 200-400 ft (61-122 m); marine glauconitic green shale and sandstone and red shale; basal 250 ft (76 m); red shale, gypsum, and anhydrite wedging out eastward in southeastern half of basin.
Jurassic(?) and Triassic	0-600 (0-183)	Red cross-bedded sandstone (Nugget Sandstone) wedging out eastward; absent in eastern half of basin.
Triassic	800-1,450 (244-442)	Red shale, siltstone, and silty sandstone; widespread thin gray limestone (Alcova Limestone) in upper part, associated with thick red sandstones; olive drab dolomitic siltstone at base in western half of basin, intertonguing eastward with red shale, dolomite, limestone, and anhydrite.
Permian	150-350 (46-107)	Tan cherty dolomite, quartzitic sandstone, green and black shale, and phosphorite beds in western half of basin, intertonguing eastward with red shale, purple limestone and dolomite, and anhydrite.
Pennsylvanian	400-700 (122-213)	Upper part chiefly white to tan sandstone; lower part interbedded sandstone, cherty dolomite, limestone, and red and green shale and claystone.
Mississippian	300-900 (91-274)	Blue-gray limestone and dolomite, coarsely crystalline in part, cavernous, cherty; some red shale and widespread sandstone near top.
Devonian	0-250 (0-76)	Upper part: yellow, gray, and pink dolomitic siltstone, shale, and dolomite; lower part: fetid brown hard dolomite and tan limestone; present only in western one-third of basin and wedges out eastward.
Silurian	Absent	
Ordovician	0-400 (0-122)	Gray dolomite, hard, massive, siliceous; forms cliffs on outcrops; gray marine sandstone locally at base; wedges out eastward at about middle of basin.
Cambrian	200-1,200 (61-366)	Widespread blue-gray marine limestone, mottled with yellow splotches; flat-pebble limestone conglomerates; green micaceous shale that is glauconitic in part, becoming sandy toward base. Widespread basal dull red sandstone in contact with Precambrian rocks.

Figure 14 shows the relation of the Wind River Canyon to the asymmetric Owl Creek Mountains with the broad gentle north flank stripped to resistant Paleozoic rocks and its steep chaotic south flank that overrode pre-early Eocene strata in the Wind River Basin. North of the canyon, the route crosses a syncline in the red Chugwater Formation (Triassic) and Lower Cretaceous rocks.

Thermopolis Overnight Stop

The overnight stop at Thermopolis is on a west-trending anticline in the Chugwater Formation. This is the site of many hot springs and hot water wells that are described in detail by Breckenridge and Hinckley (1978) and Hinckley, and other, (1982). The largest spring has a flow of about 2,900 gallons per minute of sulfur-rich water with a temperature of 133°F (56°C). Various opinions have been expressed as to whether the geothermal gradient or igneous rocks at depth, or chemical reactions are responsible for heating the water. The general consensus is that the geothermal gradient is the likely source.

East of Thermopolis is the Hot Springs State Park, on the core of an anticline in the Chugwater Formation and with good exposures of Jurassic and Lower Cretaceous strata forming hogbacks on its flanks. A paved road traverses the area and passes near several extinct hot spring vents lined with travertine. A herd of bison roams the park.

Table 3 summarizes the stratigraphic section in the Bighorn Basin. Its geologic history is presented elsewhere (Love, 1988b, p. 200-204).

DAY 2

From Thermopolis the route follows the Bighorn River for about 70 miles. The stratigraphic section for the first 10 miles is a regular sequence of Lower and Upper Cretaceous rocks dipping northward into the Bighorn

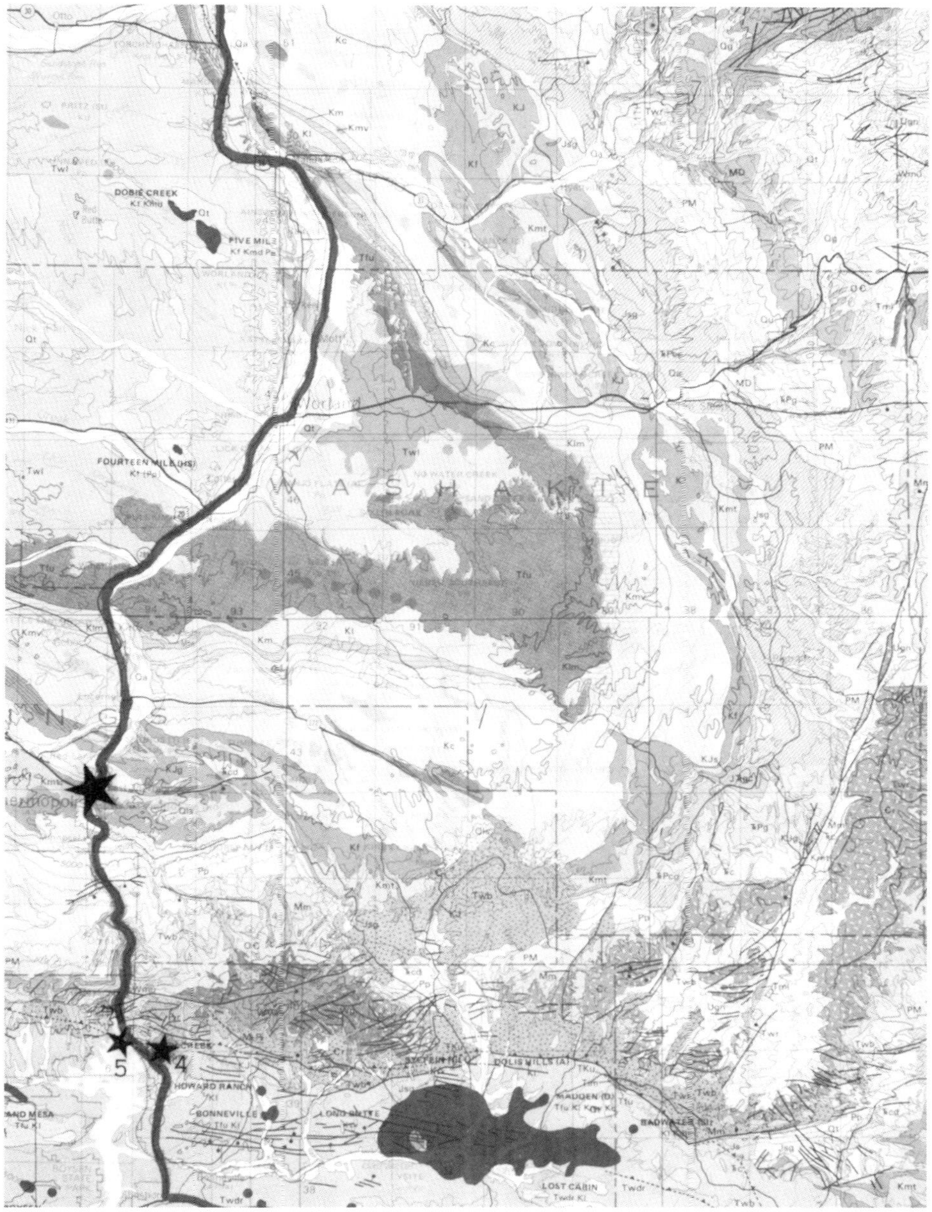

FIGURE 11 Segment of the geologic map of Wyoming (Love and Christiansen, 1985) showing the route of the field trip (heavy black line) on the last part of Day 1 and first part of Day 2. Large star indicates overnight stop at Thermopolis. Smaller numbered stars show stops along route. For geologic symbols and line patterns see source map. Squares are approximately six miles on a side. Black named areas are oil and gas fields.

Basin. The Fort Union Formation (Paleocene), consisting of drab ledgy sandstones and siltstones several thousand feet thick, is exposed in eight miles of outcrops arched up into the broad Neiber anticline. For most of the remaining distance to the town of Greybull, the highway is on alluvial valley fill along the river, but there are prominent terraces on both sides. Bedrock is chiefly the Willwood Formation of early Eocene age, contemporaneous with the Wind River Formation in the Wind River Basin.

The Pleistocene terrace sequence is complex and the chronology has been studied in detail by Palmquist (1983). He used two ash beds from eruptions in Yellowstone National Park, dated as 600,000 and about 100,000 years, to establish a time frame for some of the terraces.

Several oil and gas fields are on both sides of the highway. The largest is the Cottonwood Creek field that covers more than 80 square miles (Figure 11). Between the towns of Manderson and Greybull, the route crosses Upper Cretaceous sandstones and shales that are exposed on the flanks of a series of oil-producing anticlines.

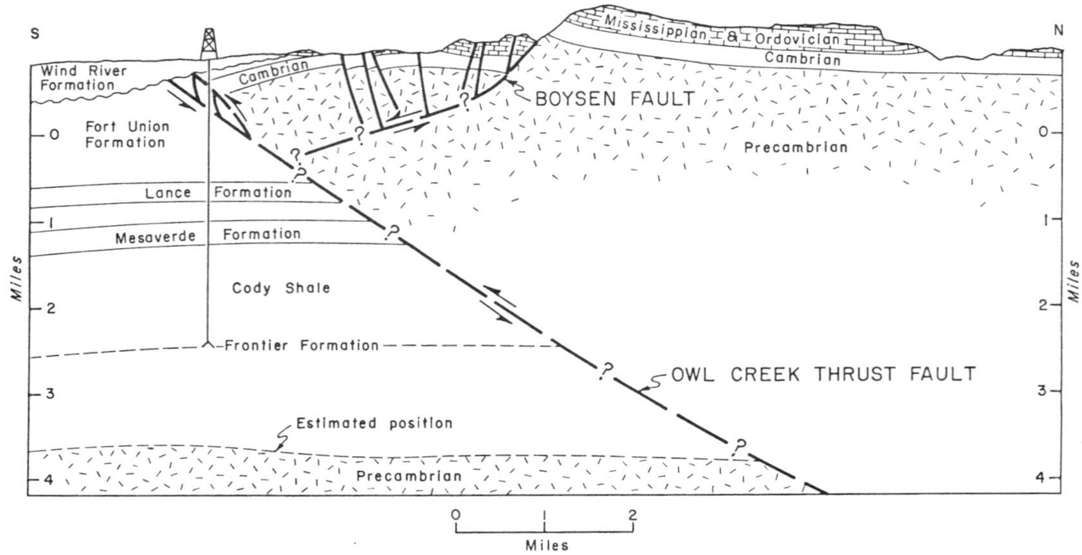

FIGURE 12 North-south simplified structure section through the Owl Creek Mountains near Wind River Canyon (after Blackstone, 1988). Gas well is located 2,700 feet northeast of Stop 4.

FIGURE 13 Oblique air photograph north across Boysen Dam and Reservoir and along Wind River Canyon. Indicated are Precambrian rocks (P€), Cambrian rocks (€u), Mississippian-Ordovician (MOu), Permian and Pennsylvanian (PPu), Boysen fault (FAULT). The lumpy hills in the lower half of the photograph are complexly folded and faulted Paleozoic and Mesozoic rocks. Photograph by P.T. Jenkins.

At Greybull, the route turns east to the west flank of the Bighorn Mountains (Figure 15) and crosses a series of anticlines and synclines, chiefly in Cretaceous rocks from which the unfolded lower Eocene basin deposits have been stripped. Three miles north of Greybull is the spectacular Big Sheep Mountain anticline with Madison Limestone (Mississippian) exposed in the core and excellent exposures of Paleozoic and Mesozoic rocks on both flanks (Figures 16, 17). A detailed structural analysis of this fold was made by Hennier and Spang (1983). An interesting and significant geomorphic feature is the course of the Bighorn River which is superimposed obliquely across the anticline through a canyon about 1,000 feet deep. Thus, it is a smaller version of the superimposed course the same river takes across the Owl Creek Mountains and, as will be seen later, across the northwestern part of the Bighorn Mountains.

Stop 6

Stop 6 is at the mouth of Shell Canyon where the stratigraphy of the lower part of the Mesozoic and all of the Paleozoic sequence can be seen. Six geologic quadrangle maps, detailing the stratigraphy and the rather complex structure of this area, were prepared by Iowa State University geology students and faculty and published in color by the Wyoming Geological Survey.

In a terrace on the north side of the Shell Creek valley near the Iowa State University geology camp is an exposure of the "Field Camp ash" of Palmquist (1983) who considers it to be about 100,000 years old so this part of the valley must have been excavated after that deposit was emplaced. Of particular interest to vertebrate paleontologists in this area are the soft puffy green to variegated claystones and siltstones in the Morrison (Upper Jurassic) and Cloverly (Lower Cretaceous) formations. These have yielded many specimens of dinosaurs. The famous Howe dinosaur quarry in the Morrison Formation, nine miles north of the village of Shell, was excavated by the American Museum of Natural History in the 1930s (Figure 18). Dinosaur sites in the overlying Cloverly Formation have been found both north and south of U.S. Highway 14 near Stop 6 (Ostrom, 1970).

Shell Creek has cut a canyon about 1,000 feet deep across the west flank of the Bighorn Mountains. Along the sides of the canyon, Ordovician, Devonian, Mississippian, Pennsylvanian, and Permian rocks are exposed in cliffs. Farther east, for about 10 miles, the highway cris-crosses

FIGURE 14 Air oblique photograph west across the Wind River Canyon in the foreground, and along the Owl Creek Mountains, showing their asymmetry. Indicated are the Grand Teton Peak in the Teton Range, 140 miles away (A), Gannett Peak in the Wind River Range (the highest in Wyoming at 13,804 feet), 85 miles away (B), the lower Eocene Wind River Formation in the Wind River Basin (C), the Washakie Needles, a dacite intrusive about 39 million years old in the Absaroka Range, 60 miles away (D), and treeless dip slopes of the Permian Phosphoria Formation on the north flank of the Owl Creek Mountains (E). Photograph by John S. Shelton.

TABLE 3 Paleozoic, Mesozoic, and Tertiary rocks within and along the margins of the Bighorn Basin. (For formation names, correlations, and areas of outcrop, refer to Love and Christiansen, 1980; 1985; after Love, 1988.)

Geologic Age	Thickness ft (m)	Description
Miocene	0-200 (0-61)	White tuffaceous sandstone and thin, locally derived conglomerate.
Oligocene	0-200 (0-61)	Light-gray, tan, and pale-green tuffaceous bentonitic claystone and siltstone and white lenticular limestone; locally derived lenticular conglomerate at base.
Late and middle Eocene	0-6,000 (0-1,830) (west side, 0-850; [0-259])	Upper part is gray and white coarse volcaniclastic conglomerate; middle part is green tuff; lower part is bentonitic variegated claystone and tuff; locally contains coal, oil shale, and uranium phosphate.
Early Eocene	2,000-3,000+ (610-914)	Claystone and siltstone, variegated to drab, interbedded with brown and gray lenticular sandstones. Tongues of locally derived conglomerate extend eastward into the basin; tongues of gold-bearing quartzite conglomerate from a distant source to the west are present near base. Thin oil shales are locally present in upper part; very little volcanic debris present.
Paleocene	1,000-10,000 (305-3,050)	Gray to buff sandstone, drab to dully variegated claystone and siltstone, and coal beds, one of which is as much as 20 ft (6 m) thick. Gold-bearing quartzite clast conglomerates in western part.
Late Cretaceous	3,000-8,000 (914-2,438)	From top down: golden-brown sandstone, white tuffaceous sandstone interbedded with thin carbonaceous shale and yellow bentonite (combined thickness about 2,500 ft; 762 m); light-gray sandstone interbedded with carbonaceous shale and some major coal beds (thickness 800-1,200 ft; 244-366 m); gray marine shale (1,500-3,000 ft; 457-1,067 m); and a basal marine sandstone, carbonaceous shale, and thin lenticular conglomerates (500-700 ft; 152-213 m).
Early Cretaceous	800-1,300 (244-396)	Marine and brackish-water black siliceous shale, bentonite beds, and tan sandstone; nonmarine sandstone and variegated claystone near base.
Jurassic	275-700 (84-213)	Uppermost part: nonmarine gray sandstone and variegated claystone; remainder marine glauconitic green shale and sandstone and red shale; basal thick gypsum beds and red shale throughout basin.
Triassic	600-820 (183-250)	Red shale, siltstone, silty sandstone; widespread thin gray limestone bed (Alcova Limestone) in upper part in southern half of basin, associated with thick red sandstones; olive-drab dolomitic siltstone at base in western half of basin, intertonguing eastward with red shale and anhydrite.
Permian	50-250 (15-76)	Tan dolomite, quartzitic sandstone, chert, and green and black shale in western part, intertonguing eastward with red shale and anhydrite. Thin phosphorite beds in southwestern part; thick gray sandstone at base along southeastern margin of basin.
Pennsylvanian	250-600 (76-183)	Upper part chiefly white to tan sandstone; lower part interbedded sandstone, cherty dolomite, limestone, and red and green shale and claystone.
Mississippian	500-1,000 (152-305)	Blue-gray limestone and dolomite, coarse-grained, cavernous, cherty; some red shale and a widespread sandstone near top.
Devonian	0-500 (0-152)	Upper part; yellow and gray dolomitic siltstone, shale, and dolomite; black shale locally at top. Lower part: fetid brown hard dolomite and tan limestone; absent in southeastern part of basin.
Silurian	Absent	
Ordovician	0-450 (0-137)	Gray dolomite and dolomitic limestone, hard, massive, siliceous; forms cliffs on outcrops; absent along southeastern margin of basin.
Cambrian	900-1,300 (274-396)	Widespread blue-gray limestone mottled with yellow splotches, flat-pebble limestone conglomerates, green micaceous shale, glauconitic in part, becoming sandy toward base. Widespread basal dull red sandstone in contact with Precambrian rocks.

some tremendous landslides in Cambrian shales. These active slides have been a constant problem in highway maintenance for many years.

At the head of Shell Creek, the route turns north onto a broad rolling upland surface that cuts across Precambrian quartz monzonite and quartz diorite about 2,900 million years old. In a few places along the highway are thin vestiges of white soft tuff, probably remnants of the White River Formation (Oligocene) which once extended all the way across the Bighorn Basin and was deposited on this upland surface (McKenna and Love, 1972, fig. 2A).

Stop 7

Stop 7 is on the Precambrian rocks at an overlook on the west side of U.S. Highway 14 (alternate) before it descends the west flank of the Bighorn Mountains. From this vantage point, the geologic panorama of the northeastern part of the Bighorn Basin can be observed. Off to the south is the Five Springs area which displays steeply dipping to overturned Paleozoic and Lower Mesozoic strata that are thrust westward toward the Bighorn Basin. Several small normal faults collapsed the steeply dipping strata mountainward from the thrust (Osterwald, 1961, fig. 4). An air oblique view of this area is shown in Figure 19.

The route descends the steep west face of the Bighorn Mountains, and in tremendous road cuts, details of the Paleozoic and Mesozoic stratigraphy and structure can be seen close-up.

Stop 8

Stop 8 provides an opportunity to see and photograph these features. Farther west, the Cretaceous strata abruptly flatten and are covered by broad terrace deposits. A few hundred yards south of the highway, on a northwest-facing point of a terrace overlooking Bighorn Lake (also called Yellowtail Reservoir) is more than 10 feet of white Lava Creek ash (the "Kane ash" of Palmquist, 1983) that has been dated as about 600,000

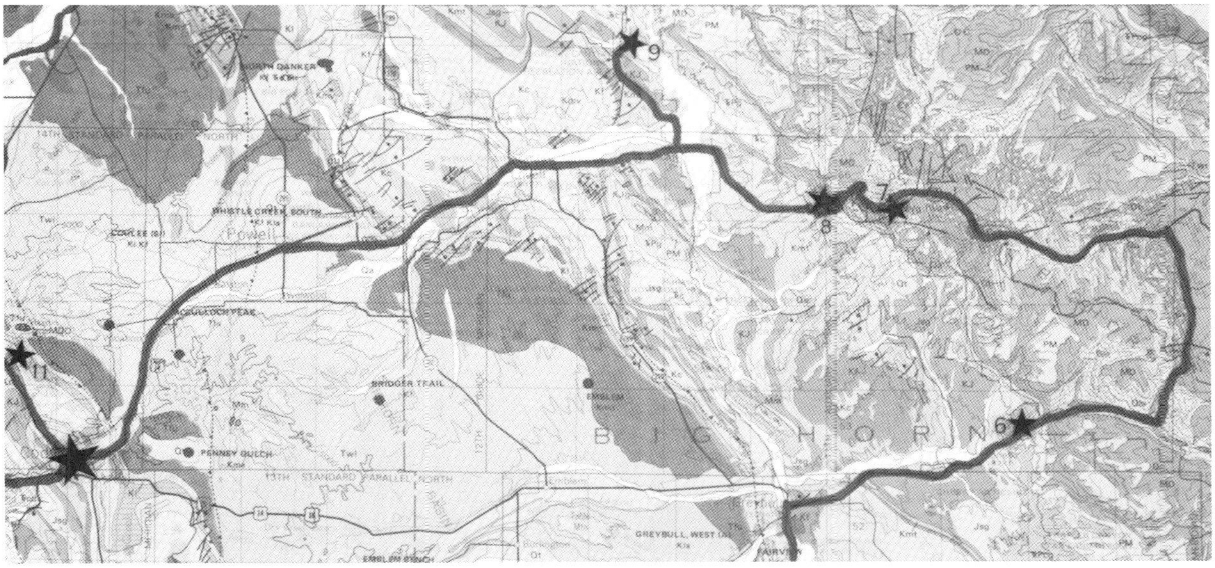

FIGURE 15 Segment of the geologic map of Wyoming (Love and Christiansen, 1985) showing the route of the field trip (heavy black line) on the last part of Day 2. Large star indicates overnight stop at Cody. Smaller numbered stars show stops along route. For geologic symbols and line patterns see source map. Squares are approximately six miles on a side. Black named areas are oil and gas fields.

FIGURE 16 Air oblique view northwest along the Big Sheep Mountain anticline. Indicated are the snow-capped Beartooth Mountains 75 miles away(A), Bighorn River superimposed across Big Sheep Mountain anticline (B), dip slope on the Permian Phosphoria Formation (C), U.S. Highway 14 (D), and Cretaceous banded Mowry shale (E). Photograph by John S. Shelton.

FIGURE 17 Air oblique view southeast along the steep plunging nose of Big Sheep Mountain anticline. Indicated are Bighorn River superimposed across the anticline (A), Triassic Chugwater Formation (B), Jurassic Sundance Formation (C) and a thick sequence of Lower Cretaceous rocks (D). Photograph by John S. Shelton.

years. Its source was a caldera in Yellowstone National Park.

Stop 9

An optional side trip to Stop 9 goes 15 miles northeast to the Devils Canyon overlook in the Bighorn Canyon National Recreation Area along State Highway 37. Waterman (1983) presents a road log of the geology along this route. The Bighorn River in the foreground at Stop 9 has cut a precipitous canyon 2,200 feet deep farther to the northwest entirely across the northern part of the Bighorn Mountains, just as it did across the Big Sheep Mountain anticline and the Owl Creek Mountains to the south. The surface on which the superimposed river course was established was probably Miocene rocks (McKenna and Love, 1970). In the foreground are red shales, dolomites, and limestones of the Permian Goose Egg Formation. The cliffs to the northeast are of Pennsylvanian, Mississippian, and Devonian rocks with a combined thickness of more than 1,500 feet (Richards, 1955). Of some geomorphic interest is the occurrence of Absaroka-type volcanic gravels 500 feet above the original river level at the mouth of Crooked Creek a mile northeast of Stop 9. If these can be accurately dated, they will provide a rate of downcutting of the river through the Bighorn Canyon.

The route west from the junction of State Highway 37 and U.S. Highway 14 (alternate) to Cody crosses the deep part of the Bighorn Basin where Precambrian rocks are at a depth of about 25,000 feet (Blackstone, 1989). In the shallower, eastern part of the basin, where the Tertiary rocks have been stripped away, a series of oil-producing anticlines in Cretaceous rocks can be seen on both sides of the route for the first 20 miles (Figure 15). The remainder of the route to within six miles of

FIGURE 18 Howe dinosaur quarry in the Morrison Formation (Upper Jurassic), nine miles north of Shell, excavated by the American Museum of Natural History in 1933. Note three men for scale. Photograph by Barnum Brown, furnished through the courtesy of M.C. McKenna.

FIGURE 19 Air oblique view north showing folded and faulted Paleozoic rocks south of the Five Springs area along the west flank of the Bighorn Mountains near Stop 8. Landslide deposits are in the lower left foreground. The high level surface at the upper right cuts across Precambrian and Paleozoic rocks. Photograph by AMOCO – Bob Lynn.

Cody crosses a complex series of Quaternary terraces (Palmquist, 1983).

Cody Overnight Stop

The town of Cody is built on a series of thick terrace gravels (Palmquist, 1983). These can be correlated with others farther east and southeast in the Bighorn Basin. The terrace deposits overlie gently dipping Upper Cretaceous rocks. The Buffalo Bill Museum-Whitney Art Gallery and Old Trail Town, both with collections of major historical interest, are located within the town limits. Two miles to the west are the De Maris hot springs with a maximum temperature of 96°F (36°C) and a flow of approximately 1,700 gallons per minute of highly sulfurous water. The springs, no longer exploited, are described in some detail, along with their chemistry, human history, and possible geothermal gradient origin of the heat by Breckenridge and Hinckley (1978). Associated with both active and extinct hot springs are thick deposits of travertine and considerable amounts of sulfur which were mined commercially in the past. The geology of the Cody area was mapped in detail by Pierce (1966).

DAY 3

The field trip route goes west from Cody on U.S. Highway 14-16. Along the river banks below the terrace gravels are good exposure of Lower Cretaceous, Jurassic, Triassic, and Paleozoic rocks down to the Precambrian granite and gneiss, on the flanks of the sharply folded and faulted Rattlesnake Mountain anticline (Figure 20). The Shoshone River has cut a superimposed course through a canyon about 3,000 feet deep, obliquely across the anticline. The highway passes through several tunnels cut in granite, gneiss, and a mafic dike complex, all of which are probably older than 2,800 million years. The highest cliffs on both sides of the canyon are Madison Limestone (Mississippian) nearly 1,000 feet thick (Figure 21). The Bureau of Reclamation outlet tunnel for water from the Buffalo Bill Reservoir encountered many large stalactites and stalagmites of sulfur in caves in the Madison Limestone. Dynamite blasts ignited the sulfur and asphyxiated several construction workers (A.G. Love, written commun., 1936).

Stop 10

This stop is at the Buffalo Bill dam, a vantage point for viewing the structure of the Rattlesnake Mountain anticline. The dam was completed by the Bureau of Reclamation in 1910, is 325 feet high, and the reservoir capacity is about 440,000 acre feet. The dam is in the process of being raised so that more water can be stored. The structure of the Rattlesnake Mountain anticline has been the subject of many papers and much speculation. The various ideas have been summarized by Blackstone (1986). He concluded that the "Rattlesnake Mountain

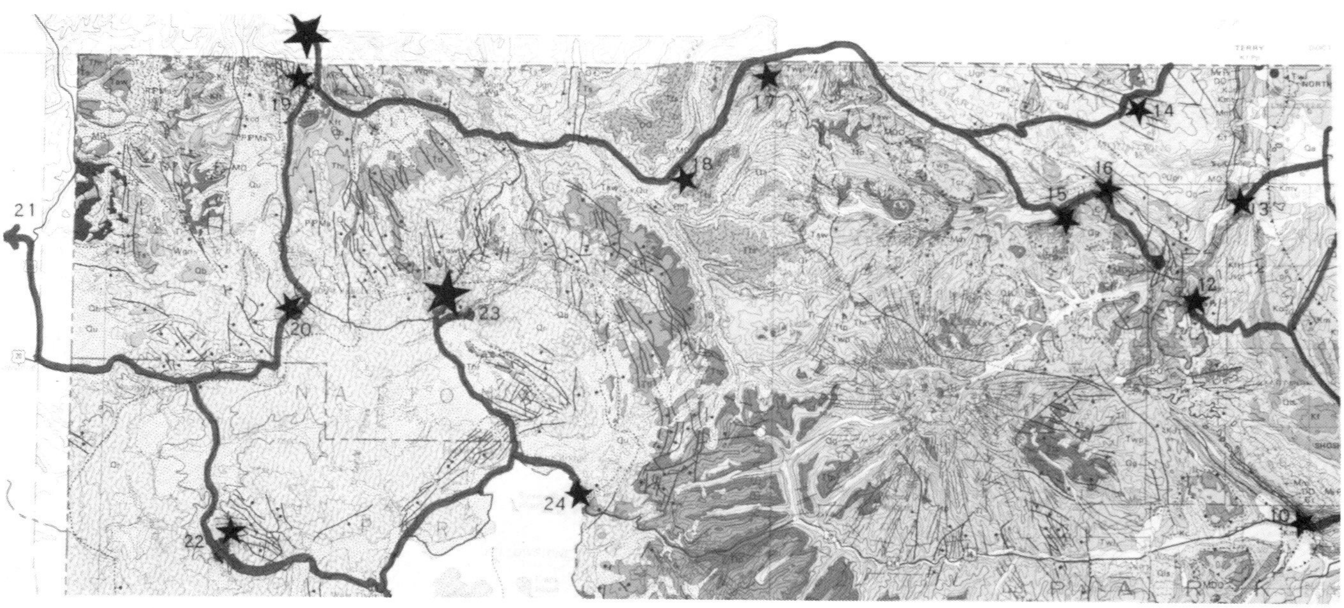

FIGURE 20 Segment of the geologic map of Wyoming (Love and Christiansen, 1985) showing the route of the field trip (heavy black line) on most of Day 3, Day 4, and Day 5. Large stars indicate overnight stop at Gardiner, Montana and Canyon (Yellowstone National Park, Wyoming). Smaller numbered stars show stops along route. For geologic symbols and line patterns see source map. Squares are approximately six miles on a side. Black named areas along the extreme northeastern margin are oil and gas fields.

FIGURE 21 Air oblique view east down the canyon cut by the Shoshone River through Rattlesnake Mountain. Buffalo Bill dam and reservoir are in foreground. The dam abutments are in Precambrian sheared granite and gneiss. The big cliff in the upper middle is Madison Limestone and the fartherest cliffs are Tensleep Sandstone (Pennsylvanian). Photograph by Jack Richard.

anticline is an excellent example of a folded surface (the interface between the Precambrian basement and the overlying sediments) modified by east dipping reverse faults, an interpretation very similar to that of Brown (1984). The structural relief on the west flank of the fold is approximately 8,000 feet (2,440 m) and has been accomplished predominantly by folding since the maximum fault displacement is approximately 2,000 feet (600 m)." Figure 22 is a view south from Stop 10, showing the interpretation of structure by Brown (1984).

Stop 11

The route retraces to Cody and then turns north on State Highway 120 for 10 miles, mostly along strike of Cretaceous sandstones and shales. Stop 11 is on Cody Shale and from here Heart Mountain (Figures 15 and 24) can be seen.

The Heart Mountain detachment fault is one of the major, and controversial, tectonic anomalies along the route of this field trip. We are greatly indebted to W.G. Pierce, who has spent many years studying and mapping this fault for the U.S. Geological Survey, for supplying the following description and photographs that document its principal features and interpretations of its origin.[1]

Because the evidence of interpreting the history for this faulting is so geographically scattered, the discussion

---

[1] Modified and updated from Pierce, 1973.

FIGURE 22 Air oblique view south at the stratigraphy and structure of the north-facing side of the canyon through Rattlesnake Mountain. Indicated are Precambrian rocks (P€), Flathead Formation and overlying Cambrian strata (€), Ordovician Bighorn Dolomite and overlying Devonian rocks (Ob), Madison Limestone (Mm), Tensleep and Amsden Formations (IPta), and Triassic Chugwater Formation (Ꞧc). Directions of fault movement are indicated by arrows. Photograph and geologic interpretation by W.G. Brown.

is presented as a single unit, related to Stop 11. Later pertinent stops along the route of Days 3 and 4 are coordinated with this presentation.

Regional Setting

The Heart Mountain detachment or décollement fault (Pierce, 1987a) is in northwestern Wyoming, adjoining and east of Yellowstone National Park. It lies on the east flank of the Rocky Mountain uplift and on the western part of the Bighorn Basin, which is both a structural and topographic basin. The Rocky Mountain uplift began in the later part of the Late Cretaceous and reached its climax during the Eocene. The Bighorn structural basin developed concomitantly with the Rocky Mountain uplift. The rocks transported by the Heart Mountain fault became detached from the eastern border of the uplift area and moved southeastward into the adjoining Bighorn Basin near the beginning of the middle Eocene.

Two smaller detachment faults, the South Fork (Pierce, 1986) and Reef Creek (Pierce, 1963), which preceded the Heart Mountain fault, are similar in movement and closely related in time to the Heart Mountain fault (Figure 25, left). They are also on the east flank of the Rocky Mountain uplift, and their movement also was to the southeast. The upper plate of the South Fork detachment was thicker and more extensive than the Reef Creek. The Heart Mountain detachment took place below the Reef Creek fault, and the scattered upper plate blocks of that fault were carried in piggy-back fashion by the Heart Mountain fault (Figure 25, right). The Heart Mountain fault rode out over the upper plate of the South Fork detachment, however, and left upper plate blocks of the Heart Mountain fault on top of the South Fork detachment. The Heart Mountain fault is the youngest and least deformed of the three faults and is the most extensive, best exposed, and best preserved.

Nature and Extent

The upper plate of the Heart Mountain fault is composed of about 490 m of predominantly carbonate rocks

FIGURE 23 Looking across Buffalo Bill Reservoir at the Sheep Mountain remnant of the Heart Mountain detachment, composed of brittle, cliff-forming Ordovician to Mississippian carbonate rocks. It lies on the South Fork detachment which preceded it, composed of ductile Jurassic to Upper Cretaceous rocks. Photograph by W.G. Pierce.

of Ordovician to Mississippian age and from 0 to 270 m or more of volcanic rocks (Cathedral Cliffs Formation) of Eocene age (Figure 26). Underlying the detachment horizon are about 365 m of Cambrian shales with some limestone beds and at the base a sandstone which rests on Precambrian granitic rocks. Volcanic rocks of the Wapiti Formation, deposited very soon after the emplacement of the Heart Mountain fault blocks, have preserved many of the structural features that would otherwise have been lost.

The Heart Mountain fault is a composite of four types of faults, indicated on Figure 27. They are (1) a high-angle breakaway fault, trending north-south for at least 37 km at the northwestern limit of the detachment fault, which developed as the detached mass moved away from strata that remained in place, (2) a bedding plane fault, 56 km long and 24 km or more wide, stratigraphically located a few feet above the base of the Bighorn dolomite, (3) a transgressive fault, (sometimes referred to as a ramp), more than 15 km long and about 3 km across, which cut obliquely up-section or transgressed the strata between the bedding fault and the ground surface, and (4) a fault along which the upper plate blocks moved up to 50 km over a former land surface.

There are more than 50 upper plate remnants of the Heart Mountain fault that range in size from a few hundred km to 8 km across (Figure 27) and many more that are from a few km to a few tens of km across. These pieces of upper plate, which before faulting covered an area of about 1,300 km^2 comprising the bedding and transgressive parts of the fault, are now scattered over a pie-shaped area of 3,400 km^2 which is about 105 km long in the direction of movement, and 13-50 km wide. The fault blocks that moved over the former land surface

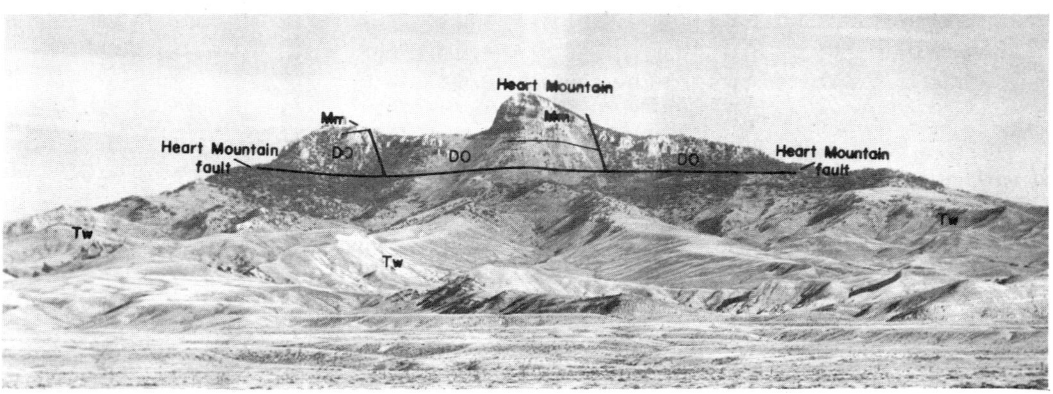

FIGURE 24 View northeast toward Heart Mountain, about three miles from Stop 11. Indicated are the lower Eocene Willwood Formation (Tw), overridden along the Heart Mountain thrust fault by the Devonian and Ordovician rocks (DO) and Madison Limestone (Mm). Photograph by Jack Richard; geology after W.G. Pierce.

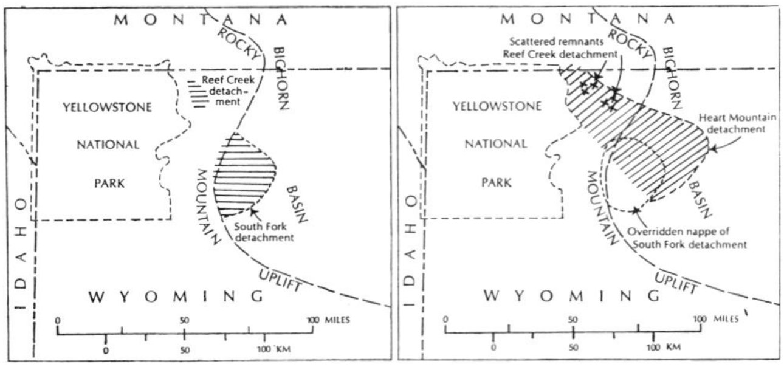

FIGURE 25 Sketch maps showing the location of the Reef Creek, South Fork, and Heart Mountain detachment faults. (left) Location of the Reef Creek and South Fork detachment faults which preceded the Heart Mountain fault. (right) Location of Heart Mountain detachment fault, and its relation to the preceding Reef Creek and South Fork Detachments. From Pierce, 1973.

were widely scattered over an area on the order of 2,100 km^2 Extensive sediment deposits composed of carbonate fragments derived from upper plate blocks attest to the former presence of larger and more numerous fault blocks on the former land surface part of the fault in the Heart Mountain area.

Although the upper plate has suffered some erosion, particularly the former land surface phase, which lacked the protective armor of volcanic rocks that covered the bedding part of the fault, the present distribution of the upper plate into scattered blocks is not due to subsequent erosion of a once-continuous sheet. The upper plate originally occupied the 1,300 km^2 area of the bedding and transgressive faults. Approximately half of that area, or about 650 km^2 was tectonically denuded and almost immediately covered by volcanic rocks (Pierce, 1987b). Thus, there was available only about 650 km^2 of the upper plate to move onto the former land surface part of the fault. The area of the land surface part of the fault, however, is about 2,000 km^2, or more than three times the area of upper plate available to cover it. To scatter 650 km^2 of upper plate over a 2,000 km^2 area requires that the upper plate be broken up into a number of smaller segments. Moreover, the lack of erosion of the surface of tectonic denudation and the preservation of carbonate fault breccia on it indicate that the upper plate broke up and separated into scattered blocks while it was emplaced.

Slope of Fault Surface

The slope on which the Heart Mountain fault blocks moved can be determined fairly closely inasmuch as the elevations and distance between the highest and lowest points on the fault can be determined fairly accurately. The southeastermost remnants moved over a former land surface that was underlain by Eocene continental beds of the Willwood Formation. The Willwood does not contain marine sedimentary rocks, so it follows that the elevation at places such as McCulloch Peaks, 21 km southeast of Heart Mountain, was at least above sea level. The distance back to the highest point, the breakaway fault, is 105 km. If a slope of 2° is projected from sea level for that distance, the elevation at the breakaway would be at least 3,700 m above sea level. Volcanic rocks at the breakaway fault, some slightly older and some slightly younger than the faulting, contain petrified wood and fossil forests. Dorf (1964, p. 9) believes these plants grew at an elevation of no more than 900 m above sea level. Consequently, a slope of less than 2° seems to be indicated.

The transgressive fault now slopes downward about 10° to the northwest, as can be observed for a distance of 2.4 km between its connection with the bedding fault on the northwest and the fault on the former land surface on the southeast. However, the dip of the transgressive fault originally may have been less than this, because there has been some postfaulting subsidence in the region northwest of the transgressive fault, which could have increased the northwest dip of the transgressive fault.

Observations on the Fault Itself

Throughout the 56 km southeastward extent of the bedding plane phase, the Heart Mountain fault occurs at precisely the same stratigraphic horizon, with only a few very minor expections. This horizon is at the top of a dolomitic limestone bed, generally about 2½ m thick, that is the basal unit of the Bighorn Dolomite (Figure 26). This is an unexpected place for a detachment or gravity type of fault to occur. The horizon at which the faulting occurred can be examined in unfaulted beds west of the breakaway fault, east of the transgressive fault, and north of the bedding fault on the flank of the Beartooth uplift. At these places the fault horizon is nothing more than an

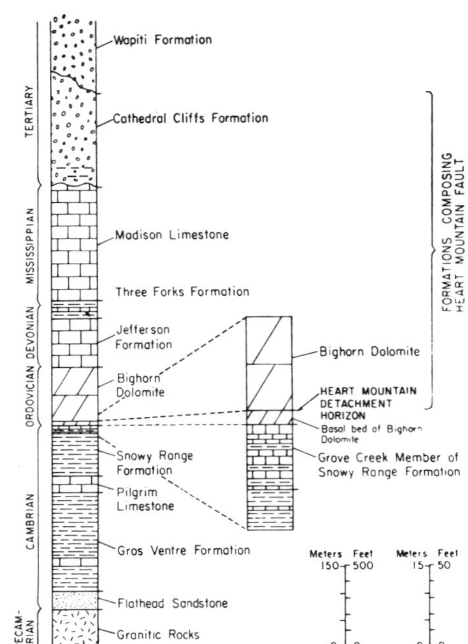

FIGURE 26 Generalized stratigraphic section showing the formations associated with the Heart Mountain fault. From Pierce, 1973.

inconspicuous bedding plane in a carbonate sequence. The faulting mechanism did not require the presence of fissile or lubricated rock – neither shale nor evaporite occur at this horizon. The mechanism did not even favor a fissile or lubricated horizon, for if it did the fault would have been slightly lower in the section, in the Snowy Range Formation or its Grove Creek Member.

Although the basal bed of the Bighorn Dolomite has some irregular vertical fractures, it is not brecciated (even immediately below the fault), unlike the beds above the fault. The strata beneath the basal bed likewise do not appear to be fractured or deformed.

Carbonate fault breccia, cemented into a hard, dense rock, rests on the fault surface at a number of places. It ranges from less than 1 cm to 1 m in thickness and exceptionally is as much as 3.6 m thick. Commonly more than 50 percent of the breccia is matrix, but the matrix is also carbonate breccia of microscopic to submicroscopic size. The megascopic breccia fragments range from specks to 2 mm or more in size; most are angular, some are subangular, and some rounded. Where the Bighorn Dolomite rests on the Heart Mountain fault, the lowermost few inches is commonly a breccia in which the fragments range in size from a few millimeters to rock flour. The breccia has the appearance of being thoroughly mixed but actually it may not be, for it begins to grade upward within inches into shattered dolomite in which the dolomite is broken into fragments similar to those below but have not rotated. Shattering has been observed to extend 15 m or more above the base of the upper plate. Breccia blocks became increasingly larger upward, and the brecciation gradually dies out. Although there is some variation in detail in the fault breccia, it is either a carbonate breccia or minutely shattered carbonate rock; it is not a mylonite. Where Madison Limestone rests directly on the fault, fault breccia is less common and slickensides may be present.

In places, the fault breccia has been injected upward as clastic carbonate dikes (Pierce 1979). A few of these clastic dikes have been noted in the carbonate rocks of the upper plate, but more of them have been found in the volcanic rocks. They are less than 1 cm to 0.6 m wide and some extend upward as much as 30 m above the Heart Mountain fault.

Time and Nature of Movement

Some of the fault blocks rest on rocks as young as early middle Eocene, and the fault is overlain in places by rocks that are either very late early Eocene or early middle Eocene. Emplacement must have occurred, therefore, near the beginning of the early middle Eocene.

The Crandall Conglomerate, a stream channel deposit of late lower Eocene or early middle Eocene age, is related to a preliminary Heart Mountain fault movement, during which roughly the northeastern one-third of the bedding and adjoining transgressive part of the fault moved less than 1 km southeastward (Pierce, W.G., and Nelson, W.H., 1973). This preliminary movement of part of the Heart Mountain upper plate occurred only a very short time before the main movement of the entire upper plate. The Crandall Conglomerate, one of many features associated with the Heart Mountain detachment that indicates tectonic denudation occurred as part of the emplacement mechanism, is briefly discussed and illustrated in the latter part of this summary.

As the upper plate moved, it broke up into many blocks which separated as they moved, and as movement progressed they became further separated. The amount of horizontal movement covers a wide range. The fragments of the upper plate nearest to the breakaway fault have moved from less than a km to a few km's. The most southeasterly remnants of the plate have moved some 50 km at least, from the transgressive fault. Nowhere did the upper plate imbricate or ride up on itself. The fault blocks were most widely separated on the former land surface, as indicated in Figure 27. In the transgressive fault and adjoining area to the northwest, the fault blocks are less scattered. Presumably, the scattered blocks that were moving along the bedding fault were retarded as they started moving up along the transgressive fault. As additional blocks moved down the bedding fault they would abut against the slower moving blocks, and some of their forward momentum would be transferred to the blocks moving up the transgressive fault.

The main movement occurred during a brief interval of time, as shown by (1) the absence of erosion on the surface of tectonic denudation in the bedding fault area, which was exposed during faulting, and by (2) the clastic carbonate dikes. The carbonate dikes are composed of Heart Mountain fault breccia that was left on the surface of tectonic denudation as the fault blocks separated; the carbonate fault breccia was injected into volcanic rocks of the Wapiti Formation that subsequently filled gaps between the blocks. This finely divided carbonate material could not have survived lengthy exposure to erosion. Consequently, the Heart Mountain fault movement is believed to have been very rapid, probably cataclysmic.

The Emplacement Mechanism

The presence of a breakaway fault rather than a root zone and the separation of the upper plate into numerous blocks clearly indicate that the Heart Mountain fault was not an overthrust emplaced by a push from the rear. Once the detachment or shearing off had started along the bedding and transgressive faults, the force of gravity could have accounted in considerable part for the continuing movement of the upper plate, but some factor other than gravity definitely seems to be needed to explain each of the following characteristics of the fault:

1. The slope of less than 2° on which the fault blocks moved seems to be much too low for movement by gravity alone, particularly on a fault that lacks any sign of being lubricated. In some places today, as along the

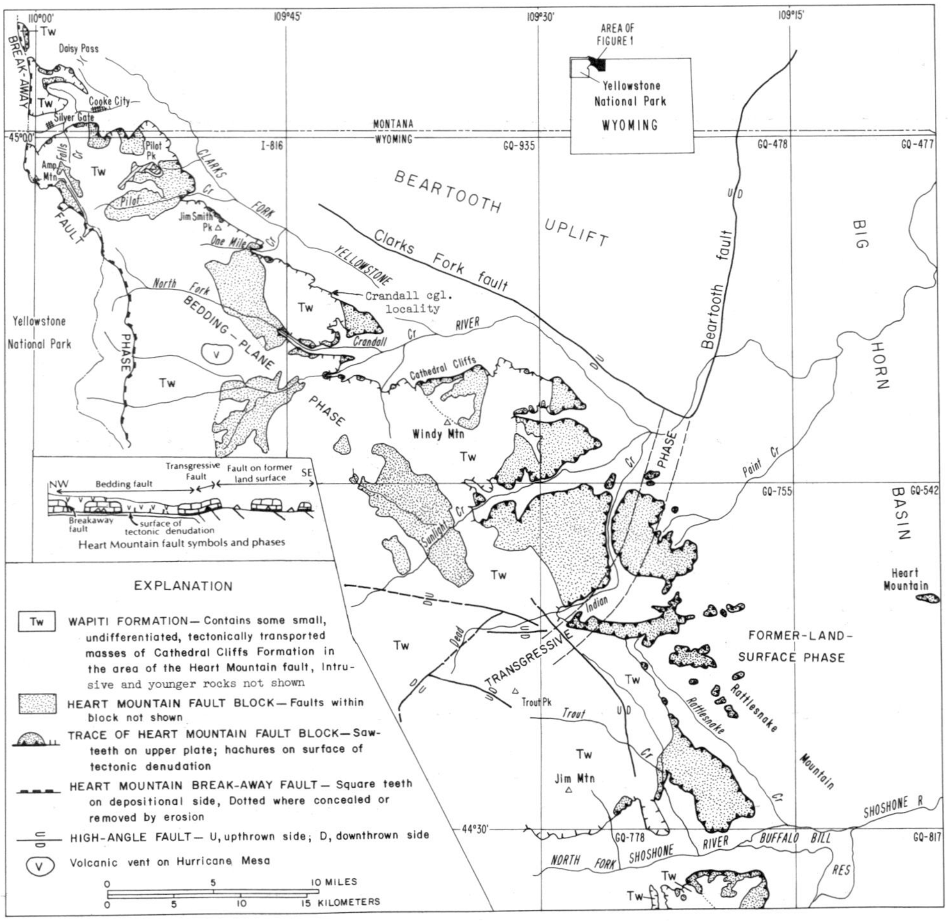

FIGURE 27 Map showing location of four phases of Heart Mountain fault, the location of the fault blocks, and the distribution of the Tertiary Wapiti Formation lying between and on them. Modified from Pierce, 1987b. GQ numbers refer to Geologic Quadrangle Maps and the I number to a Miscellaneous Investigations Series Map published by U.S. Geological Survey at a scale of 1:62,500.

transgressive fault, upper plate blocks rest on the Heart Mountain fault where it slopes as much as 10°, but they are not now sliding under the influence of gravity.

2. The horizon along which the Heart Mountain bedding fault occurred was in the most viscous rather than in the least viscous strata. If the bedding fault was due primarily to gravity sliding, however, then it should have been in the soft Cambrian shales less than 15 m below the detachment horizon, for they have a lower viscosity than the dolomite and dolomitic limestone in which the deformation occurred. Indications are lacking in the Cambrian shale of movement within them, or of stratigraphic displacement or angular discordance, whereas at the fault horizon between the basal Bighorn bed and the overlying dolomite (Figure 26) evidence of stratigraphic displacement, angular discordance, and fault breccia are common features.
3. The fact that the entire area of the bedding fault is at the same stratigraphic horizon, and also that the surface of tectonic denudation exposed by the bedding fault was not eroded, seems to indicate that the initial movement on it occurred essentially simultaneously over either the entire fault surface or over a large part of it, a conclusion that in either case carries important implications regarding the amount of shear stress that is required. According to Hubbert and Rubey (1959, p. 125), a fault produced by stress applied to the rearward edge of the upper plate most likely would be propagated as a dislocation in which the stress involved at any one time affects only a minute fraction of the total area of the fault surface, and therefore can be assumed to be negligible. In the Heart Mountain bedding fault, however, deformation did not take place as a result of stress applied to the rearward edge, so the foregoing conclusion is not applicable. If shearing simultaneously occurred over all parts of a bedding fault area of several hundred square miles, a great deal of shearing stress would be required. That shearing stress would be in addition to the force needed to keep the plate moving once motion had started.

The unspecified factor that seems to be indicated by the three items just discussed would be the one that operated in combination with the constant force of gravity to produce both lateral transport and dispersal of the parts of the upper plate. In particular, the need for a mechanism that would have the effect of reducing friction along the fault by at least several orders of magnitude has led to six suggested mechanisms for the Heart Mountain fault. Three of them are similar in one respect: friction is greatly reduced owing to the presence along the fault of fluid, either liquid or gas, under pressure approaching the lithostatic pressure of the upper plate. These are the fluid pressure mechanism of Hubbert and Rubey (1959; also Rubey and Hubbert, 1959), the air-cushion landslide mechanism suggested by Hsu (1969), and the hovercraft (volcanic gas) mechanism of Hughes (1970). Gravity sliding by shear in low-viscosity strata has been proposed by Kehle (1970) and gravity spreading by Hauge (1985). Pierce (1973) has proposed earthquake oscillations.

Earthquake Oscillation

A combination of gravity and earthquake oscillations is the only mechanism that has been suggested (Bucher, 1947, p. 196; Pierce, 1963, p. 1234) which is compatible with the observed features of the Heart Mountain fault. The common association of volcanism and earthquakes, and the close association in time between the extrusion of the Wapiti volcanics and the faulting, point up the likelihood of there having been earthquakes at that time. The vertical displacement of 6,100 m along the east front of the Beartooth uplift, most of which occurred during Early Tertiary time, no doubt was accompanied by earthquakes (Pierce, 1963, p. 1234). The cataclysmic nature of the Heart Mountain fault movement, which is suggested by the short time involved in emplacement is consistent with an earthquake-associated mechanism.

If under certain unusual conditions a unit of gently inclined rocks received slightly more upward acceleration than a unit of rocks on which it rested, the frictional resistance to movement between them would be lessened and the units would tend to separate along an incipient fault. If, during the upward cycles of oscillatory seismic motion, accelerations approaching 1 g were imparted to the rocks above the incipient fault, then, at the moments that the upward accelerations ceased, the stress normal to the fault as a result of gravity would approach zero and separation would tend to occur at the base of the rock unit with higher acceleration. In effect the upper plate would be almost weightless. With innumerable repetitions of upward acceleration, the upper plate would be intermittently nearly unrestrained by friction and free to move laterally on a very low slope. Moreover, perhaps a forward impetus might be imparted to the upper plate by seismic waves that emanated from beneath the upper plate and moved tangentially upward to the southeast.

It has been stated that the maximum acceleration of the ground within 8 km of the epicenter of a magnitude 5.0 shock may be 5-10 per cent gravity, and that a magnitude 6.0 shock can have a maximum ground acceleration on the order of 20 percent of gravity (Natl. Research Council, 1969). During the San Fernando earthquake of February 9, 1971, accelerations much higher than these were recorded. Thus in at least some instances, previous ideas as to the maximum acceleration of earthquakes may be revised upward. The San Fernando earthquake, of moderate magnitude 6.6, has, according to Maley and Cloud (1971, p. 165), provided the largest collection of significant strong-motion data ever obtained. They report (p. 163):

"At Pacioma Dam, 8 km south of the epicenter, the earthquake accelerations were the highest ever recorded, i.e. in the 0.5 to 0.75 g range with several high-frequency peaks to 1 g. The initiation of strongest shaking began 1 to 2 seconds after instrumental triggering and lasted for approximately 12 seconds."

Degenkolb (1971, p. 133) comments in regard to apparently the same record:

"It is reported that the closest strong motion record of this earthquake, taken at or near the summit of a fractured rock formation, indicated ground accelerations of more than 100 per cent g horizontal and 70 per cent g vertical. For 12 seconds the recurring motions were recorded at about 50 per cent g. This would indicate that ground motions for even a moderate earthquake may be much greater than assumed by many engineers. . . . It is interesting to note that these high concentrations of force, motion, or energy are on some form of rock rather than on saturated alluvium."

In discussing records obtained in the upper levels of buildings, Maley and Cloud (1971, p. 164) report that "Peak values in excess of 0.4 g were recorded as distant as downtown Los Angeles, nominally 42 km from the epicenters, although the vast majority of measurements were in the 0.2- to 0.4-g range. . . . The highest top-floor accelerations are approximately two to three times those measured at the ground level in the same structure."

In discussing the vertical accelerations at Kagel Canyon, Morrill (1971, p. 177) reports that the San Fernando earthquake "produced evidence of vertical accelerations far in excess of any ever recorded instrumentally." Evidence at the Los Angeles County Fire Station No. 74, 13 km from the epicenter, ". . . suggests that the vertical acceleration exceeded that of gravity and may have equaled or exceeded that of the great Assam earthquake of June 12, 1987 in very local situations. . . . A 20-ton fire truck enclosed in a garage move 6 to 8 feet fore and aft, 2 to 3 feet sidways without leaving visible skid marks on the garage floor. The truck was in gear, and the brakes were set. . . . Marks which appear to have been made by the right rear tire were found on the door frame 3 feet above the floor, while the metal fender extending several inches out beyond the upper portion of the tire was not damaged. Four feet above the floor the hose rack was broken by the rear step of the truck. The step was bent up while the hose rack was broken downward."

The living quarter's building has shingle siding which

overlapped the foundation 4-3/8 in. After the quake a corner of the building was displaced 11 in. laterally from its former position. To reach that position the building must have been elevated more than 4-3/8 in. relative to its foundation before it was displaced laterally. This "suggests that the building accelerated upward, with respect to the ground, at a rate of at least 1 g for about 0.1 second; that is, if one assumes only 1 g and the given distance."

An unusual effect of the San Fernando, California, earthquake of February 9, 1971, is the shattered earth at Wallaby Street, Sylmar, which furnishes some information regarding the effects of seismic waves (Nason, 1971). Some of the special features of the shattered earth site which Nason has described in detail are:

". . . the tossed-earth phenomenon of overturned soil, which may indicate a seismic shaking exceeding 1.0 g; the pulverization of the soil may mean that the high accleration occurred over many wave cycles. . . . The very localized nature of the high acceleration indicated by the tossed-earth may result from unknown effects of constructive interference of seismic waves arriving along different paths and possible focusing or concentration of seismic wave energy at the narrowed-down ridge-top. This suggests that seismic wave interference effects are probably very important to the detailed distribution of strong shaking and earthquake damage. . . . The fact that the soil was thoroughly overturned is an indication of the strength of the shaking. It would appear that possible seismic wave interference and focusing patterns may be very important to the local distribution of strong seismic shaking and related damage."

The purpose in citing this unusual effect is to suggest that some as yet unknown kind of focusing and combining of seismic waves may have occurred in the region beneath the Beartooth Mountain uplift so as to produce the initial shearing along the bedding fault and aid in the lateral transport of the upper plate. In this connection it is noted that the Heart Mountain fault is essentially at a seismic velocity discontinuity since the velocity of the compressional or P wave in the Bighorn Dolomite is considerably higher (possibly 50 percent or more) than in the underlying Cambrian shale.

The lack of fault striations along the Heart Mountain detachment also should be taken into account in considering the mechanism for emplacement. Pierce (1987b) suggests they are lacking for two reasons. (1) Brecciation increases downward in the basal part of the Paleozoic allochthons and grades into carbonate fault breccia of varying thicknesses, from less than a centimetre to 3.5 m. Brecciated carbonate rock that grades into a carbonate rock-flour fault breccia on the fault surface is not likely to have striations. Likewise striations are not likely to form on the fault surface of an allochthon of volcanic rocks that has brecciation and fault breccia at its base. Hauge (1985, p. 1445) noted that he could not find striae at the base of any of the allochthons of Paleozoic rocks or at the base of the allochthon of Cathedral Cliffs volcanic rocks north of Pilot Creek. (2) The Heart Mountain fault allochthons moved on a slope of less than 2 degrees (Pierce, 1973). For downslope movement to occur on a slope that low, the effect of downward pressure exerted by gravity on an allochthon at the moment of lateral movement must be reduced many-fold. As movement did occur, it can be assumed that the effect of the gravitational force on each allochthon was minimal at the moment lateral movement occurred. At the moment of movement, a low component of effective vertical pressure was exerted at the fault surface, and striae were not formed. The absence of mylonite anywhere along the detachment fault also indicates low downward pressure at the moment of movement.

The Crandall Conglomerate, A Feature Indicating Tectonic Denudation[1]

The Crandall Conglomerate (Figure 28), a stream channel deposit formed as a result of a preliminary detachment movement, is one of many features indicating tectonic denudation by the Heart Mountain detachment (Pierce, 1987b). The conditions under which it was deposited and emplaced reveal that tectonic denudation began with the first movement that took place on the Heart Mountain fault.

The Crandall Conglomerate more than 100 m thick, is believed to have formed as a result of preliminary movement of the Heart Mountain fault. Initial movement of that fault opened a gap within upper-plate rocks in which

FIGURE 28 The regional setting and channel-like nature of the Crandall Conglomerate (Tc) as seen looking southwest from Red Lodge-Cooke City highway to Hurricane Mesa on skyline. Eocene volcanic rocks of the Wapiti Formation (Tw) rest on the Heart Mountain detachment (HM). The rocks below are the Snowy Range (€s), Pilgrim Limestone (€p), Gros Ventre and Flathead Formations (€gf), and Precambrian granite (p€g). Photograph by W.G. Pierce.

[1]Modified from Pierce and Nelson, 1973 and Pierce, 1987.

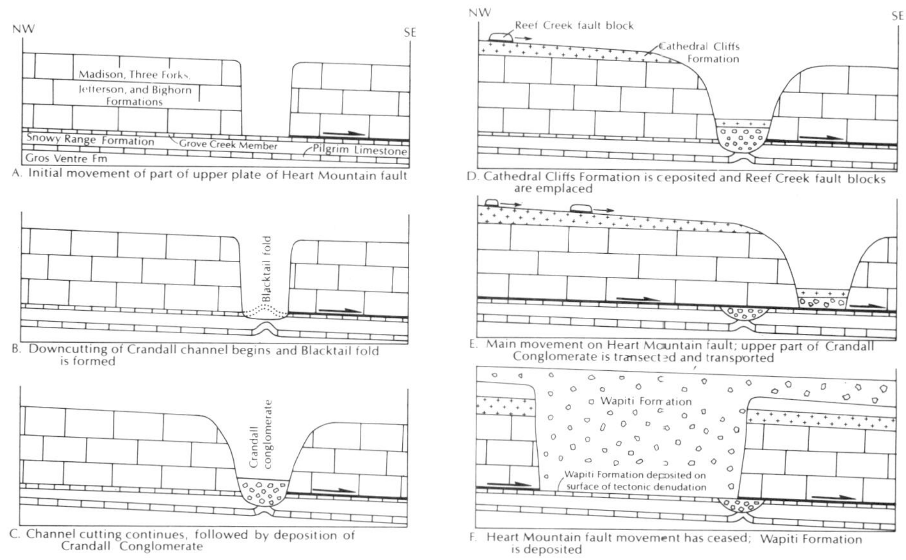

FIGURE 29 Diagram showing sequence of events associated with the Crandall Conglomerate, from A (older) to F (younger). From Pierce and Nelson, 1973.

the conglomerate was later deposited (Figure 29A). The tectonically denuded area within the rift was less than 1.6 km wide and was bordered by 600-m cliffs, mostly of Paleozoic limestone. Before the gravel was deposited, unconfined Cambrian shale below the rift was deformed into the Blacktail fold (Figures 29B and 30), a sharp anticline having no apparent roots, while streams carried away the upwelling shale and cut a channel several hundred feet deep. The debris that accumulated in this channel is the Crandall Conglomerate (Figure 29C). Deposition of the conglomerate was followed by Cathedral Cliffs volcanism and emplacement of the Reef Creek fault blocks (Figure 29D). Subsequent main movement on the Heart Mountain fault (Figure 29E) left the lower part of the conglomerate in place but carried the upper part some 24 km southeastward with rocks of the upper plate. Sheared cobbles at the top of the Crandall Conglomerate in the lower plate (Figures 31 and 32) and upper-plate rocks that rest on conglomerate in-place in the lower part of the channel (Figure 33) clearly indicate that the Crandall Conglomerate was distributed by the Heart Mountain fault movement, not prior to it as stated by Hauge (1985, p. 1453). Most of the Crandall deposits of the lower plate rest directly on the Blacktail fold. Of the

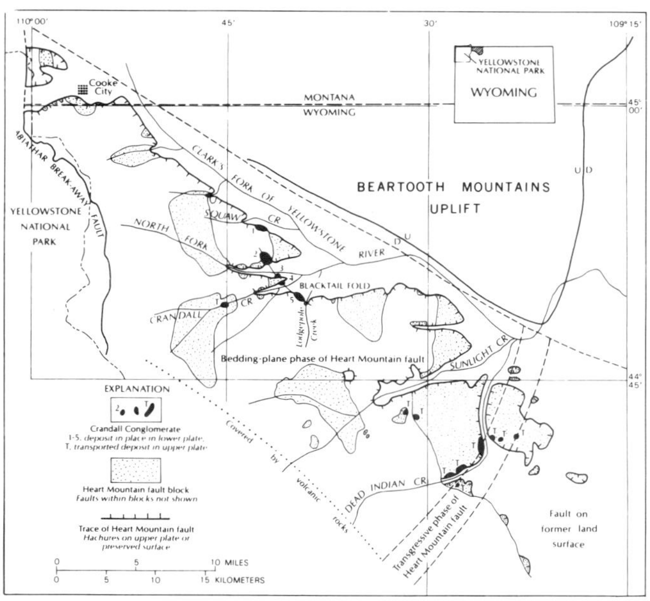

FIGURE 30. Map showing distribution of the Crandall Conglomerate in relation to Blacktail fold and Heart Mountain blocks. From Pierce and Nelson, 1973.

FIGURE 31 Crandall Conglomerate (Tc) sheared by the Heart Mountain fault. Heart Mountain fault breccia (HMb) overlies the fault. This site is marked on Figure 28. From Pierce and Nelson, 1973.

15 known deposits of Crandall Conglomerate, five are in place but have been overridden by the upper plate of the Heart Mountain fault, and ten have been transported as part of the upper plate (Figure 30). After this movement, volcanic rocks of the Wapiti Formation blanketed the region (Figure 29F) as shown by Wapiti that overlies sheared Crandall Congomerate at the top of the lower plate (Figures 28 and 31).

The Blacktail fold is the only known fold in the 300-km^2 area of the Heart Mountain fault. A stream channel deeply incised precisely along its crest can best, and perhaps only, be explained as due to a preliminary movement that affected a segment on the border of the bedding-plane part of the Heart Mountain fault. That

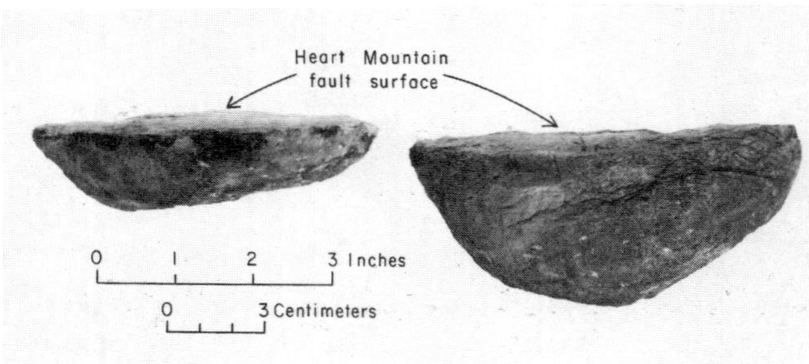

FIGURE 32 Sheared limestone cobbles taken from the upper surface of the Crandall conglomerate shown in Figure 31. The Heart Mountain fault, which passed horizontally across the outcrop, sheared many of the cobbles in half, and their upper parts became part of the upper plate. Small offsets along shears in cobble on right are toward the southeast, which is the direction of movement on the Heart Mountain fault. From Pierce and Nelson, 1973.

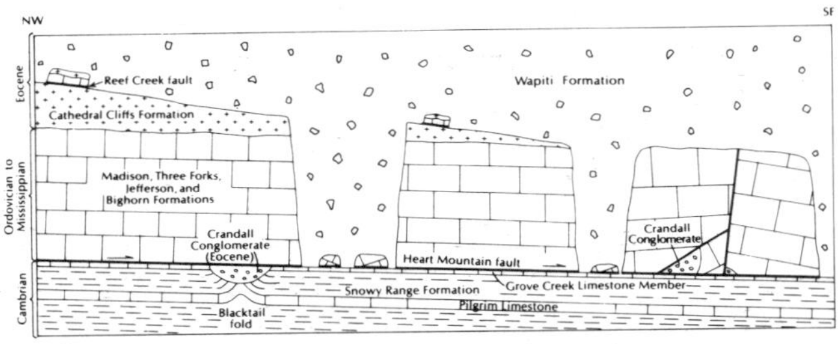

FIGURE 33 Diagrammatic cross section showing the relation of the Crandall Conglomerate in the lower and upper plates to the rocks and deformation with which it is associated. An 8-ft-thick (2.4 m) basal bed of Bighorn Dolomite, too thin to be shown in this diagram, overlies the Grove Creek Limestone. From Pierce and Nelson, 1973.

movement produced an initial break-away fault that was a forerunner of the much larger one 15 to 25 km to the west (Figures 27 and 34).

A deposit of Crandall Conglomerate on Crandall Creek (Figure 30) lies on, and is part of, an allochthon of Paleozoic rocks. Hauge (1985) mentioned this deposit as an "unresolved complication" for his "coeval-with volcanism" model because he assumed that the distribution of the Crandall Conglomerate predated the Heart

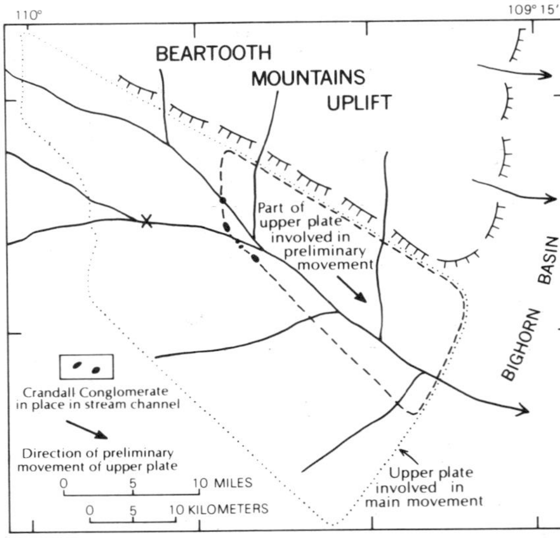

FIGURE 34 Sketch map showing preliminary movement in early Eocene time of a segment of the upper plate of the Heart Mountain fault and the deposits of Crandall Conglomerate that are in place. Conglomerate originally deposited near X was moved about 5 mi (8 km) southeastward to its location on Crandall Creek (Fig. 30) by the main Heart Mountain fault movement. From Pierce and Nelson, 1973.

Mountain fault. The deposit, however, does not present an unresolved complication for the tectonic-denudation model because the part above the Heart Mountain fault plane (Figure 29E) has clearly been transported by the Heart Mountain fault (Pierce and Nelson, 1971; Pierce and others, 1973).

After leaving the Heart Mountain Stop 11, the route continues northwestward along a strike valley in the Cody Shale (Upper Cretaceous). The Heart Mountain anticline to the east is marked by ridges of sandstone in the Mesaverde Formation. Oil is produced from the Frontier Formation in this anticline. About eight miles northwest of Stop 11, the route turns sharply to the west on the Dead Indian Hill secondary highway (State Highway 296) and switch-backs up the hill for a vertical rise of 3,000 feet, and crosses excellent exposures of rocks from Upper Cretaceous to Pennsylvanian (Table 4).

Stop 12

Stop 12 at Dead Indian Pass (elevation 8,048 feet) the road skirts the northeastern margin of part of the Heart Mountain thrust plate that put Madison Limestone on Triassic and older rocks. Figure 35 shows the panorama to the northwest of this stop. Indicated are various blocks of Paleozoic strata of the Heart Mountain thrust plate and their relation to older Paleozoic and Precambrian rocks of the Beartooth Mountains. The timing and mechanics of emplacement of this thrust plate are included in the discussion by W.G. Pierce of the Heart Mountain fault at Stop 11.

A noteworthy feature of this panorama is the absence

TABLE 4 Stratigraphic section of rocks along and east of the Beartooth Mountains (after D.U. Wise, 1982, *in* Foose, and others, 1986).

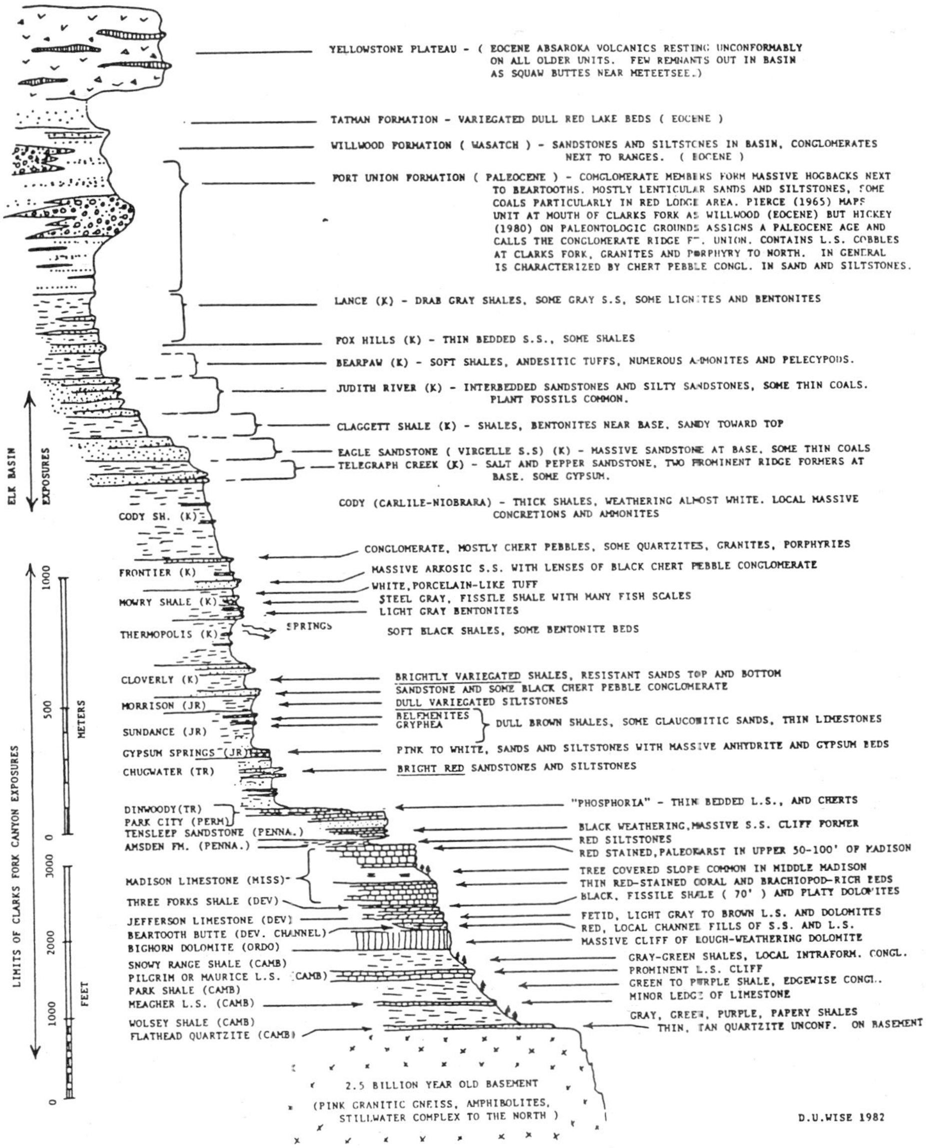

of trees on the Cambrian strata in contrast with the abundance on other rocks. There are many prehistoric Indian camp sites along Sunlight and Dead Indian Creeks, and extensive archeological excavations are in progress. Some of the occupations are older than 6,000 years. Apparently the open, grassy plains on the Cambrian strata, coupled with the abundance of water and wildlife, made this an attractive place for the prehistoric Indians.

The retracing of the route down Dead Indian Hill and back to State Highway 120 affords another view of the excellent stratigraphic section. Lateral equivalents of part of it can be seen again from Stop 13.

The route continues north for about 12 miles to the junction with State Highway 292, then follows the latter west for about 11 miles to the mouth of Clark Fork Canyon at Stop 13.

Stop 13

The stratigraphy along this part of State Highway 292 has been described elsewhere (W.G.A., 1983, p. 239-241), and has been mapped by Pierce (1965). The steeply dipping lumpy brown hills on both sides of the road are conglomerates mapped by Pierce as part of the Willwood Formation (Table 4) of early Eocene age. Dutcher, and others (1986), however, consider them probably Paleocene. They were derived from the Beartooth Mountains block to the west as it rose and was thrust eastward during the Laramide Orogeny. One of the hills about five miles northeast of Stop 13 rises about 500 feet above the valley floor and is capped by several klippen of conglomerate, probably gravity slide blocks emplaced during the latter stages of conglomerate deposition.

Stop 13 is at the end of the pavement at the mouth of

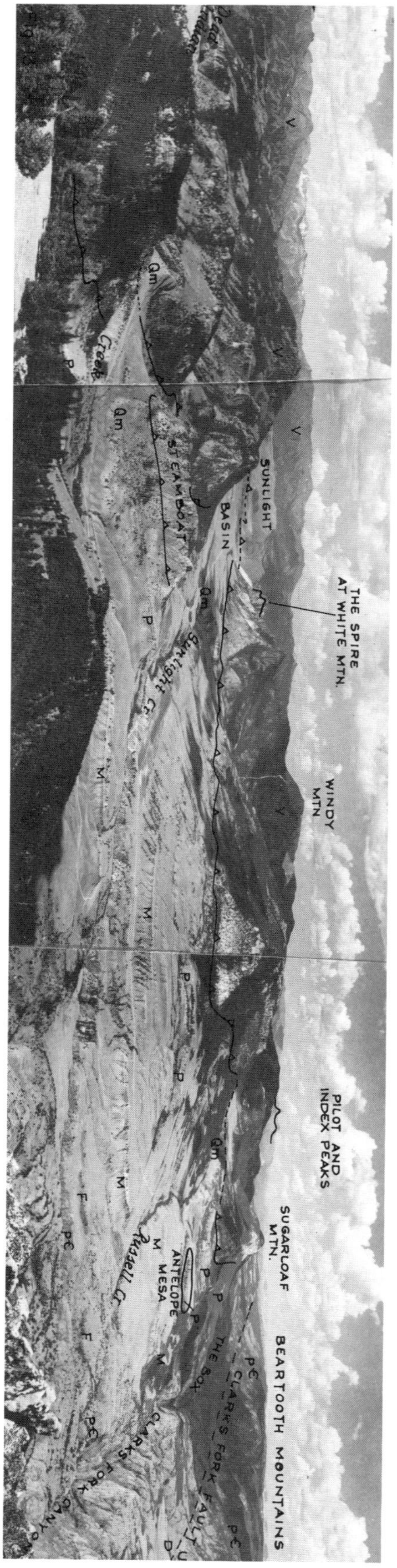

Symbols: Qm, glacial moraine; V, volcanic rocks; P, Pilgrim Limestone; M, Meagher Limestone; F, Flathead Sandstone; p€, Precambrian.

FIGURE 35 View from Dead Indian Hill summit looking northwest from near Stop 12 along the bedding plane phase of the Heart Mountain fault (barbed line). The tree-covered hill in left center foreground is a fault block in the trangressive fault phase of the detachment that passes into the land surface phase at the summit. The canyon of the Clarks Fork is 1000 feet deep. Photograph and annotations by W.G. Pierce.

Clark Fork Canyon. Figure 36 is a view northeastward down Clark Fork Canyon showing outcrops of Precambrian granite gneiss whose age is about 2.7 billion years, overlain unconformably by Cambrian, Ordovician, Devonian, and Mississippian strata in this view. Of interest to glaciologists is the multi-storied valley carved by several advances of glaciers. These have not been studied in detail but lateral and terminal moraines representing at least two glaciations can be seen along the canyon and in front of the canyon mouth. Evidence of a very old high level glaciation is present on the Beartooth Plateau at an elevation of 9,700 feet on the north side of the canyon, at the eastern lip of the plateau. It is interesting to speculate as to what happened when the ice, advancing southeastward across the plateau from Beartooth Butte, 14 miles away, poured off the edge of the Beartooth Mountain front into the Bighorn Basin, 5,000 feet below (if the topography was about the same then as now).

Figure 37 is a view of the structure on the south side of Clark Fork Canyon. The thrust fault is not shown on most of the geologic maps of this area, and apparently does not involve the stratigraphic section north of Stop 13. That section appears to be complete from the base of the Cambrian to the Mesaverde Formation of Late Cretaceous age. Details of the structure along this part of the mountain front have been described by Wise (1983), and a cross section is presented by Blackstone (1986, Pl. 2, sec. D-D′).

An unique stratigraphic occurrence in the section north of Stop 13 is the presence of fossil fish at the top of the red Chugwater Formation. Fossil fish are rare in this part of the Mesozoic sequence. These are not well preserved but have been described by Meyer (1984, p. 21-22) as possibly *Lepidotes* sp. with a range from Late Triassic to Early Cretaceous. Meyer considered them to be in an unnamed lens of red, maroon, and tan pebbly siltstone as much as 20 feet thick, of uncertain age, but probably in the basal part of the Middle Jurassic Gypsum Spring Formation.

The route returns to State Highway 120 and continues north to Red Lodge, Montana, a total distance of about 45 miles. The geology along this route has been described in detail by Foose, and others (1986, p. 291-306). A map of the regional geology has been published by Foose, and others (1961, Pl. 1). The route is chiefly on Quaternary deposits and the Fort Union Formation. There are eight major terrace levels along the streams draining the Beartooth Mountain front. One the basis of incorporated ash, the second highest is dated at about two million years and the fourth from the youngest at about 600,000 years (Reheis, 1987). These are especially prominent along Rock Creek, downstream from the town of Red Lodge.

Red Lodge, Montana, Overnight Stop

The human history of the Red Lodge area has been described at length in a 410-page book "Red Lodge – a

FIGURE 36 Air oblique view northeast down Clark Fork Canyon showing Archean granite gneiss probably more than 3,400 million years old. Several episodes of glaciation have made the canyon multi-storied. Near the mouth of the canyon and at the upper right are Cambrian, Ordovician, Devonian, and Mississippian strata. The bottom of the canyon in the foreground has an elevation of 4,700 feet and Bald Peak at the upper right is at 8,630 feet. Photograph by Jack Richard.

saga of a western area" (Zapan and Owens, 1979). Coal mining in the Fort Union Formation was the chief industry, other than livestock. Some of the mines were extensive and operated for many years. During World War I coal production was more than a million tons annually, but after a series of mine disasters and competition from other areas, all mining ceased.

From the standpoint of modern geology, the most interesting event in Red Lodge history was the establishment of the Yellowstone-Beartooth Research Association (YBRA) in 1936, at a camp three miles west of Red Lodge. This organization continues to be active and in the last half century has had an illustrious history of scientific research as well as the training of some of the most talented geologists in the world (Bonini, and others, 1986, p. 1-9).

DAY 4

From Red Lodge westward the route is along U.S. Highway 212 (the Cooke City Highway). Directly west of Red Lodge is the steeply-dipping Linley Conglomerate, of uncertain age but derived from the Beartooth Mountains. It rests unconformably on the Fort Union Formation. In the vicinity of the YBRA field camp, a section of steeply-dipping Cambrian to Mississippian rocks is exposed. West of it is the main Beartooth thrust fault along which the Precambrian rocks override the northwestern margin of the Bighorn Basin. In this area, Blackstone (1986, Pl. 1) estimated the Tensleep Sandstone to be at a depth of about 15,000 feet below the eastern lip of the thrust fault. Hypothetical cross sections based in part on gravity profiles are presented by Bonini and Kinard (1983, p. 93).

After crossing the thrust, the route continues west on the Precambrian core of the Beartooth Mountains and climbs along the spectacular highway from 5,500 feet to 10,936 feet at the Summit Overlook.

Stop 14

Figure 38 shows the view northward toward Twin Lakes and the ice-scoured terrain that has modified the Beartooth Plateau. Glaciers have cut precipitous canyons 2,000 to 3,000 feet deep, chiefly in pink granite gneiss. It

FIGURE 37 Air oblique view south near the mouth of Clark Fork Canyon south of Stop 8, showing the stratigraphy and structure of the east front of the Beartooth Mountains. Photograph and annotations by W.G. Brown.

FIGURE 38 View north from Beartooth Highway near Summit BM 10,936 feet, showing the glaciated Twin Lakes surrounded by Archean granite gneiss which is probably more than 3,400 million years old. Barely visible at far left is down-faulted gray monzonite porphyry of Late Cretaceous age. The rolling subsummit surface cut chiefly across Precambrian rocks has a general elevation of about 10,000 feet. Photograph by Jack Richard.

has been dated as about 2,700 million years old. Some of the other gneisses may be older. The gray to buff monzonite at the extreme left margin of Figure 38 is part of a series of intrusive rocks of Cretaceous age.

Farther west the route goes over summit after summit, culminating at Beartooth Pass (11,029 feet). Intensely glaciated terrain lies to the north, contrasting with the rolling Beartooth Plateau to the south. About 12 miles west of this pass is Beartooth Butte, a remnant of nearly horizontal, but somewhat faulted Cambrian, Ordovician, and Devonian strata rising 1,700 feet above the adjacent terrain. This is an important site because (1) these Paleozoic rocks are the source of the glacial boulder train previously described on the plateau edge north of Clark Fork Canyon, and (2) the red limy sandstone lens on the southeast face of the butte is the type section of the Beartooth Butte Formation of early Devonian age. This lens contains museum-quality specimens of primitive bony ostracoderms and also some fossil plants. The presence of Beartooth Butte and other nearly flat-lying islands of Paleozoic rocks at nearly the same elevations as the Precambrian rocks on the Beartooth Plateau suggests that this part of the plateau is a stripped surface, much of which was shaped before the first Cambrian strata were deposited on it, and has hardly been folded in the last 500 million years. About seven miles west of Beartooth Butte, the route doubles back sharply to the southeast on State Highway 296, along the Clark Fork of the Yellowstone River.

Stop 15

Three miles southeast of the highway junction, the Crandall Conglomerate of Eocene age is exposed in a cliff face on the south side of the river. For details of the regional setting, local details, geologic history, and significance of the Crandall Conglomerate, refer to Figures 27-34 and the discussion by W.G. Pierce of the Heart

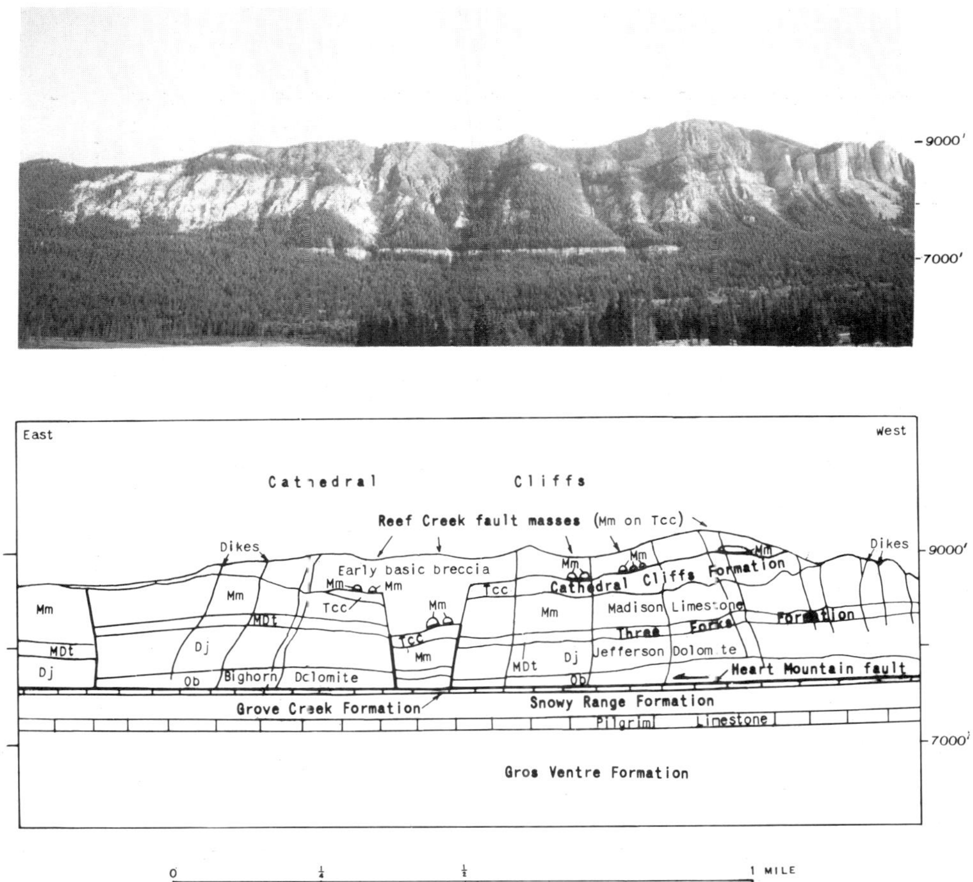

FIGURE 39 View and section looking south at face of Cathedral Cliffs, showing a large detached Heart Mountain fault block of Paleozoic carbonate formations and Cathedral Cliffs volcanic rocks overlain by Reef Creek detachment blocks of Madison Limestone, that have been carried piggyback as part of the Heart Mountain detachment. The "Early basic breccia", now Wapiti Formation, was deposited after movement ceased, followed by lateral emplacement of closely spaced igneous dikes, of which only a few reach the fault and none extend below it. Photograph and diagram by W.G. Pierce.

Mountain thrust fault at Stop 11. This is one of the critical places to observe the relation of the fault blocks to the conglomerate and to the younger volcanic rocks.

Stop 16

Stop 16 is about eight miles farther southeast along State Highway 296, opposite Cathedral Cliffs which can be seen to the south across the river. Figures 39 and 40 show a panorama of this section of the Heart Mountain thrust fault and details of its structural relations to overlying and underlying strata. The geologic history and significance of structural details shown here are discussed by W.G. Pierce in the narrative on the Heart Mountain thrust fault at Stop 11.

The route backtracks to U.S. Highway 212 and continues west for about 14 miles to Cooke City, Montana. All along the north side of the highway is the rolling terrain of Precambrian rocks of the Beartooth Mountains. On the south side the higher peaks are part of the Eocene volcaniclastic and igneous rocks of the Absaroka Range that originally lapped out across the Beartooth plateau and later were stripped back. The lower slopes just south of the river are chiefly flat-lying Cambrian rocks directly below the nearly horizontal Heart Mountain thrust plane. The two sharp spires two miles west of the highway are Index Peak (the northern one, 11,319 feet) in the Wapiti Formation, and Pilot Peak (11,708 feet) in the Trout Peak Trachyandesite. The rocks on both are of middle Eocene age.

The town of Cooke City, Montana, is flanked on the lower slopes by Cambrian strata that have been intruded and overlapped by Eocene igneous and volcaniclastic rocks. This is the area of the "New World Mining District"

FIGURE 40 Nearer view of block of Madison Limestone (Mm) which was emplaced on the Cathedral Cliffs Formation (Tcc) by the Reef Creek detachment prior to the Heart Mountain detachment, and then carried piggyback as part of the latter detachment. The Reef Creek and Heart Mountain detachments are overlain by Wapiti volcanic rocks (Teb) Photograph and annotation by W G. Pierce.

FIGURE 41 Grey-colored Wapiti volcanic rocks (Eocene) overlying the light-colored basal bed of Bighorn Dolomite (Ordovician), Cambrian shales, and cliffforming Pilgrim Limestone. At the time the Paleozoic rocks overlying the basal Bighorn were removed by the Heart Mountain detachment movement, the basal Bighorn was exposed and the Wapiti was deposited on that surface. A residue of all-carbonate fault breccia unmixed with volcanic rock at the contact between the Wapiti and Bighorn at many places indicates the Wapiti was depositionally emplaced on a tectonically denuded surface and the Heart Mountain detachment was a catastrophic event. Photograph by W.G. Pierce.

described by Lovering (1929). The district was first exploited in the 1870s and mining continued for the next 75 years. There is a variety of types of ore deposits, the products being chiefly copper, silver, gold, platinum, and lead. Most are clustered around quartz monzonite and syenite stocks. The mines have been inactive in recent years. The route continues westward across the drainage divide that separates the southeastward-flowing Clark Fork and the southwestward-flowing tributaries to the main Yellowstone River. The route goes west of Cooke City for about three miles and passes through the village of Silver Gate. About a mile farther west is the northeast entrance to Yellowstone National Park.

Stop 17

Figure 42 is a view south from the northeast entrance and shows the actual break-away fault where the Heart Mountain detachment originated. West and southwest of here the Paleozoic rocks are in an unfaulted normal sequence. Details of the evolution of the Heart Mountain fault and its relation to overlying volcanic rocks and underlying Paleozoics as demonstrated at this site are given in the discussion by W.G. Pierce at Stop 11.

For the remainder of Day 4 and parts of Days 5 and 6, the following maps and publications are useful in understanding the stratigraphy, structure, thermal features, and geologic history of Yellowstone National Park: U.S. Geological Survey Map (1972) covering the entire Park, Smedes and Prostka (1972) for the Eocene volcanic rocks,

FIGURE 42 View from the northeast entrance to Yellowstone National Park looking south along the break-away fault (B) which terminates downward at the surface of tectonic denudation and the bedding horizon of the Heart Mountain detachment (HM). Post faulting Wapiti volcanic rocks (Y) on the left were deposited on both the surface of tectonic denudation and the break-away fault during the very brief time they were exposed. Slightly older volcanic rocks, Cathedral Cliffs and Lamar River (O) on the right, overlie Paleozoic carbonate rocks (P). The continuous cliff below them is the Pilgrim Limestone, 90 m below the detachment. Photograph and annotation by W.G. Pierce.

Christiansen and Blank (1972) for the late Cenozoic rhyolites, Ruppel (1972) for the pre-Tertiary rocks in the northern part of the Park and Love and Keefer (1975) for those in the southern part, Pierce (1979) for glaciation in the northern part of the Park, and Keefer (1972) for the general geology of the Park. Details of local features are referenced in the field trip narrative at pertinent places.

From Stop 17, the route continues west for a mile, then turns southwest along Soda Butte Creek, past Barronette Peak which rises 3,000 feet above the valley floor (Figure 43). The lower ledges are nearly horizontal Cambrian, Ordovician, Devonian, and Mississippian carbonate rocks overlain by likewise horizontal volcaniclastic strata of the Lamar River Formation which is about 49 million years old (Smedes and Prostka, 1972). The top 800 feet of the peak is the Wapiti Formation,

FIGURE 43 View west at Barronette Peak rising 3,000 feet above the valley floor of Soda Butte Creek. The Paleozoic sedimentary rocks at the base of the exposures are Cambrian, Ordovician, Devonian, and Mississippian carbonates. Most of the remainder is the Lamar River Formation consisting chiefly of andesitic volcaniclastic strata about 49 million years old. The top 800 feet is the Wapiti Formation, another volcaniclastic sequence about 48 million years old. U.S. Geological Survey photograph.

about 48 million years old. Both of these formations have previously been observed along the south side of Clark Fork River near Stops 15 and 16.

Stop 18

Stop 18 is at Soda Butte (elev. 6,655 feet), a white travertine cone on the east side of the highway. This is the farthest northeast evidence of thermal activity in the Park. From this stop, the appearance of the volcaniclastic rocks of the Lamar River and Wapiti Formations can be seen on Druid Peak (elev. 9,583 feet) to the northwest and Mount Norris (elev. 9,936 feet) to the southeast. Both have Madison Limestone at their bases. To the southwest, across the Lamar River is the famous "Fossil Forest" in the Lamar River Formation (Figure 44). These petrified trees have been known and discussed by many researchers since the Territorial Survey of 1872. Fritz (1982) reviewed the literature of the preceding 100 years, measured six detailed stratigraphic sections, and presented an interpretation of the torrential type of fluvial deposition that overwhelmed the trees with coarse water-laid volcanic debris. Some of the trees are rooted, whereas others have been rafted in by powerful streams. A middle Eocene rodent, tentatively identifed as *Paramys* sp. has recently been found at the base of the big stump indicated in the upper middle of Figure 44.

From Stop 18, the route goes west down the Lamar River (Figure 45) to where it joins the Yellowstone River and from there to the overnight stop at Gardiner, Montana. It is approximately along the interface between the Precambrian rocks of the Beartooth Mountains and the overlapping volcanic rocks, chiefly of Eocene age. Some of the volcanics, however, are as young as Quaternary (U.S.G.S., 1972). The route passes through Mammoth on the way to Gardiner, four miles farther north. The geology of this section is described on the return trip at Stop 19.

Gardiner, Montana, Overnight Stop

The geology of the Gardiner area has been mapped and described by Fraser, and others (1969) and the regional geology by Foose, and others (1961). The U.S.G.S. geologic map of Yellowstone National Park (1972) extends north as far as Gardiner. The panorama of geology in each direction from the town of Gardiner is very different: (1) to the northwest are huge quarries in faulted Pleistocene travertine which is as much as 40 feet thick and is used for decorative stone; (2) north and northeast are five flows of late Cenozoic basalt; (3) the higher part of the Beartooth Mountains to the north and east are Precambrian schist and quartzite and to the east-southeast, gneissic granite; (4) to the southeast the gneissic granite has been thrust southwestward during the Laramide orogeny over a thick section of Upper Cretaceous gray sandstones and shales on Mount Everts, along the Gardiner thrust fault; (5) to the south and southwest is a vast series of landslides involving Cretaceous shales; and (6) the high peak to the southwest is Sepulcher Mountain

FIGURE 44 Giant petrified tree stumps in the middle of the Specimen Ridge section of the Lamar River Formation. These trees are rooted. The arrow at the base of the big stump in the upper center of the photograph marks the site of a fossil jaw, probably of the middle Eocene rodent *Paramys* sp. National Park Service photograph.

FIGURE 45 Relief map of Yellowstone National Park, Grand Teton National Park, and adjacent areas showing the relation of the field trip route to major mountains, basins, drainages, and lakes. Large stars indicate overnight stops. Smaller stars show stops along route.

(elevation 9,652 feet), rising more than 4,000 feet above the town, and is composed of andesite breccias of the Eocene Sepulcher Formation.

DAY 5

Stop 19

The route retraces the last five miles of Day 4 to Mammoth Hot Springs. These have been described in many publications during the last 118 years. Annotated bibliographies have been compiled elsewhere (Breckenridge and Hinckley, 1978; Bargar, 1978). The springs have deposited the largest area of travertine in Yellowstone National Park, covering most of a square mile. The water cascades down about 400 feet, across a series of cliffs and pools (Figure 46), some brightly colored by algae. The total discharge is about 900 gallons per minute of clear water with a temperature of 158°F (70°C). There are no geysers in the area. Most of the water disappears underground and is thought to mingle with the much larger (10,000 gallons per minute) and cooler (128°F, 53°C) Hot River Spring along the Gardiner River 1.5 miles to the northeast.

White, and others (1975) described a drill hole at the Mammoth Springs that encountered 253.5 feet of travertine before going into Jurassic (?) strata. A maximum temperature in this hole measured 163°F (73°C). A disturbingly large amount of radon gas has been measured in some of the Park Service buildings at Mammoth; the immediate source appears to be the travertine. Additional investigations are being made.

Hot springs have been depositing travertine in and adjacent to the Mammoth area for a long time. Terrace Mountain, one to two miles southwest of the active hot springs, is capped by more than 200 feet of cliff-forming travertine with a thorium-uranium age of 63,000 ± 9,000

FIGURE 46 View northeast showing part of the travertine terraces of Mammoth Hot Springs. Yellowstone National Park Headquarters buildings are in the middle distance and peaks of Precambrian rocks of the Beartooth Mountains are on the skyline. Wyoming Travel Commission photograph.

years (Rosholt, 1976). These springs have been inactive for thousands of years (Bargar, 1978). The entire area was glaciated by the most recent ice advance (Pinedale glaciation of Pierce, 1979; see northeast corner of the Frontispiece in his publication, showing ice scour lines crossing from Swan Lake Flat over Terrace Mountain).

About two miles south of Mammoth Hot Springs, Bunsen Peak (elevation 8,564 feet) rises more than 2,000 feet above the hot springs. It is an elliptical-shaped intrusive of dacite more than a mile across and has a fission track age of 47.6 ± 1.9 million years (L.L. Love, and others, 1976).

Three miles east of Mammoth Hot Springs is Mount Everts (elevation 7,841 feet), rising about 1,800 feet above the Gardiner River. On the west face of the mountain is a well-exposed, uncomplicated section of several thousand feet of Lower and Upper Cretaceous gray sandstones and shales. Fraser, and others (1969) measured and described these strata (as had Brown, 1957) and named two new Upper Cretaceous formations; the Everts, unconformably overlain by the Landslide Creek. The relation of these new named units to those designated by Brown here and to those mapped in the southern part of the Park, 55 miles to the south, and still farther south in Jackson Hole are shown by Love and Keefer (1975, Figure 9). The Everts is apparently equivalent to the Sohare Formation in Jackson Hole (Love and Christiansen, 1985) and the Landslide Creek to the Harebell Formation of Jackson Hole and the Lance of the Bighorn Basin (Table 4).

Mount Everts is capped by the Huckleberry Ridge Tuff, an ash flow welded tuff of latest Pliocene age (2.1 million years). Two to four miles farther east is the Gardiner thrust (or reverse) fault which extends for 25 miles or more along the southwest margin of the Beartooth Mountains (Fraser, and others, 1969; Foose, and others, 1961). This fault movement occurred during the Laramide orogeny at or near the end of Cretaceous time, and put Precambrian rocks on Mesozoic; it is complicated by many imbricate blocks. The Eocene volcaniclastic rocks are not involved with the thrusts but are with later generally parallel Tertiary and Quaternary normal faults.

From Mammoth the route goes south between Bunsen Peak and Terrace Mountain, affording a close view of both the dacite and the travertine. A big highway cut at Golden Gate shows details of the flow structures in the Pliocene Huckleberry Ridge Tuff. The highway emerges southward onto the broad Swan Lake Flat which has been scoured by northeastward-moving ice (see Frontispiece in Pierce, 1979, for regional panorama of this scouring). The highway is along the edge of the Norris-Mammoth fracture zone which is about 22 miles long and consists of a series of north-south-trending faults, rhyolite and basalt vents, gravity and magnetic anomalies, and active thermal features. About two miles to the west is the Gallatin horst, consisting of upthrown Percambrian, Paleozoic, Mesozoic, and volcanic rocks in the Gallatin Range.

Directly east of the highway for several miles is the gray Swan Lake Flat Basalt of Quaternary age, overlapping the Lava Creek Tuff. This tuff is composed of rhyolitic welded ash flows with an age of 600,000 years.

Obsidian Cliff, on the east side of the highway at Crystal Springs, represents the Roaring Mountain Member of the Plateau Rhyolite, and has an average age of about 171,000 years (R.L. Christiansen, written commun.). Most of the obsidian here has many vugs and crystals but some is of such good quality and unflawed that it has been quarried extensively by prehistoric Indians. It is generally assumed that the best obsidian so widely distributed in the western United States come from this locality, but actually there are many other sites with just as high quality obsidian in northwestern Wyoming. People who stop and look at the obsidian along the highway are disappointed to find it not up to their expectations.

The remainder of the route to Norris Geyser Basin is flanked by cliffs of the Lava Creek Tuff. Roaring Mountain on the east is a spectacular, active thermal area. During the last 120 years of observation, the intensity of fumarole activity has diminished as is shown by photographs. This is a hydrothermal explosion area with the conduits coming up through the Lava Creek Tuff. From here to Norris Geyser Basin at Stop 20, the hydrothermal activity and noticable alteration of the bedrock increases. Most of the major explosive events are thought to have occurred in late Pleistocene time.

Stop 20

The Norris Geyser Basin is the site of Stop 20 and demonstrates the variety and intensity of past and present thermal activity in Yellowstone National Park. A magnificent report on this area is by White, Hutchinson, and Keith (1988). Hutchinson is one of the local experts on this trip. The importance of this area is best summarized by the following quote (White, and others, p. 2):

"Norris Geyser Basin, for many geologists, geochemists, and park visitors, is the most interesting and diverse thermal area in Yellowstone Park. The reasons, supported in detail in this report, are:

1. The basin includes most varieties of hot springs, geysers, fumaroles, and mud volcanoes found elsewhere in the park. Only travertine ($CaCO_3$)-depositing springs are absent. Steamboat Geyser (>100 m high at peak of major eruptions) is the park's highest erupting geyser of historical record; Black Growler fumarole (as hot as 138°C, Allen and Day, 1935) has the highest recorded surface temperature; and research drill hole Y-12 has the highest temperature yet measured in the park – 237.5°C at 331.6 m depth (White and others, 1975) – and also has the highest temperature yet recorded for its depth in world geothermal systems (excluding volcanoes and directly related volcanic fumaroles).
2. Norris Basin displays a wide variety of chemical types of waters, precipitates, and altered rocks, including nearly the total range found elsewhere in the park.
3. The area demonstrates the longest history of thermal activity of any of the major geyser basins. Initial activity, although not well dated, predates all deposits of the Pinedale stages of glaciation (about 12,000 to 115,000 years B.P.) and probably exceeds 150,000 years B.P. In contrast, the Firehole River geyser basins provide no evidence for activity older than early Pinedale glaciation.
4. Changes in activity in Norris Basin occur more frequently and often more vigorously than in other Yellowstone thermal areas. Change is characteristic of all high-temperature sinter-depositing geothermal systems, but is especially characteristic of Norris Basin, probably related to its exceedingly high subsurface temperatures and the highest SiO_2 contents of all thermal waters in the park."

After leaving Norris Geyser Basin, the route goes down the Gibbon River Canyon, with the upper part of the Lewis Canyon Tuff on the west and the Gibbon River rhyolite flow, about 90,000 years old (R.L. Christiansen, written commun.) on the east. About three miles south of the beginning of the canyon, in a road cut on the east side, is a localized deposit of stratified white pumiceous clay about 20 feet thick, containing mats of white, partly silicified stems of rushes and other plants and abundant black carbonized twigs and stems. The carbonized plants have a ^{14}C age of 9,440 ± 300 years (Meyer Rubin, written commun., 1964, USGS Lab no. W-1364). Diatoms are likewise abundant. This deposit is apparently related to one of the explosive hydrothermal events, probably in the Norris Geyser Basin area.

The next part of the route is down the Madison Canyon (the Madison River begins at the junction of the Gibbon and Firehole Rivers at Madison Junction), along the north margin of the concealed Yellowstone Caldera fault (Christiansen and Blank, 1974). At Madison Junction, the Nez Perce Creek rhyolite flow (about 160,000 years old) comes in from the southeast, the West Yellowstone rhyolite flow (118,000 years old) from the southwest, and various units in the Lava Creek Tuff from the north.

Along both sides of the Madison Canyon there is evidence of rock falls from the Hebgen earthquake of 1959. This earthquake had an intensity of 7.1 (Richter scale) and was felt throughout an area of 600,000 square miles. Since that time, the seismic activity in and near

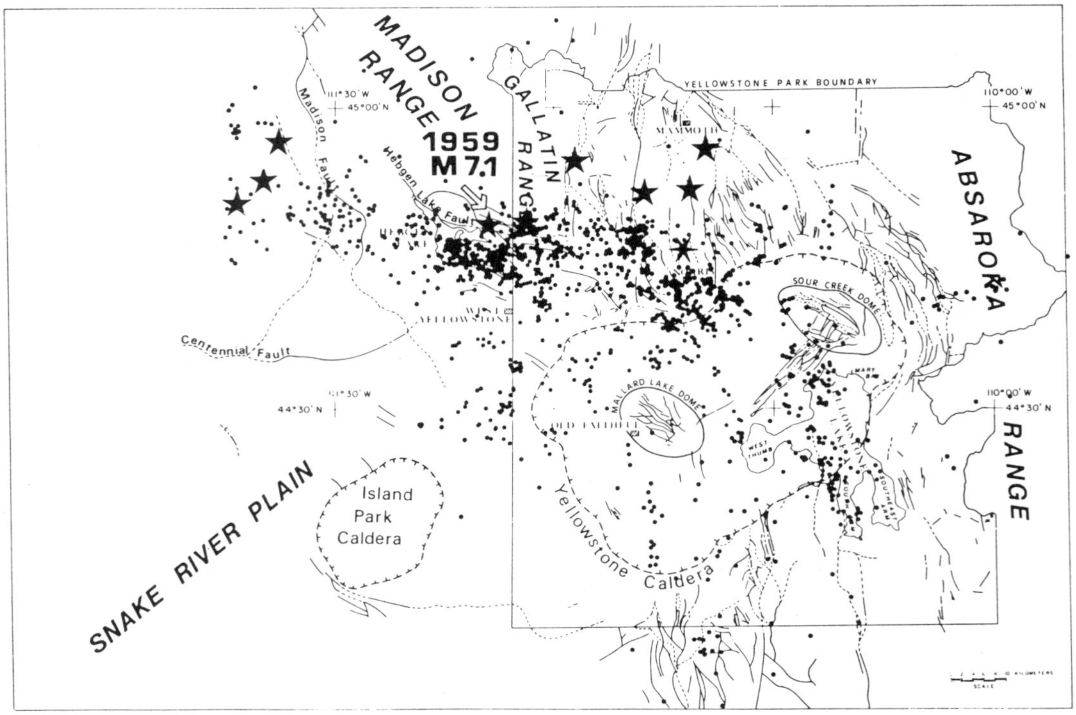

FIGURE 47 Earthquake distribution of the Yellowstone-Hebgen Lake region (1971-1979). Epicenters are those accurately determined from data of the University of Utah and the U.S. Geological Survey. Faults are taken from a compilation by R.L. Christiansen (U.S. Geological Survey, personal communication, 1978) Calderas are shown corresponding to their respective ages: I, 2.0 m.y.; II, 1.3 m.y.; and III, 0.6 m.y. Large stars correspond to epicenters of magnitude 6 to 7.1 historical earthquakes. After Smith and Braile (1984).

Yellowstone National Park has been closely monitored and swarms of epicenters of various intensities have been located (Figure 47).

A detailed geologic map from Madison Junction all the way to the Hebgen slide is in Professional Paper 435 (1964, Pl. 5; in part, Figure 20, this report). After leaving Madison Canyon, the route turns north at the town of West Yellowstone, Montana, on U.S. Highway 191 and travels across an immense alluvial dark gray obsidian sand plain extending to and north of the Madison River. About four miles north of the river the highway crosses the Duck Creek fault scrap which was formed at the time of the Hebgen earthquake (Figure 48). Thirty years later the scarp, which was 6-12 feet high, has melted down so completely that is is no longer recognizable as a fault scarp and looks more like an ancient stream terrace.

Directly north of Duck Creek the route turns northwest on State Highway 499-287 and follows the edge of Hebgen Lake. Duck Creek scarp is the first of many related to the Hebgen earthquake that can be seen on the north side of the highway for the next 15 miles. One of these is about two miles northwest of Duck Creek where a fault cut across the Blarneystone ranch buildings (Figure 49) with devastating consequences. In addition to the normal faults, there is a series of much older, Laramide parallel thrust faults with Mississippian to Cambrian strata thrust northeastward over Permian and Triassic rocks. The young normal faults cut the overriding plates of the older thrusts.

From the Hebgen Dam the highway turns southwest along the shore of "Quake Lake" and is on Quaternary deposits until it climbs up onto the landslide dam overlook. The higher hills on both sides are Precambrian rocks of the Madison Range.

Stop 21

Stop 21 is at the overlook at top of the landslide dam. U.S. Geological Survey Professional Paper 435 consists of 20 papers, maps, and photographs dealing with all aspects of the Hebgen earthquake of August 17, 1959 (U.S.G.S., 1964).

The Hebgen landslide was emplaced in a matter of seconds or minutes in response to a 7.1 (Richter) earthquake. A U.S. Forest Service crowded campground at the base of the slide was obliterated and at least 28 people were killed or missing. The shock was felt throughout an area of 600,000 square miles. The maximum fault displacement was 22 feet and measurable changes in ground elevation occurred in at least 200 square miles. A section of the mountain on the south side of the canyon, 2,200 feet long, slid a vertical distance of 1,300 feet into the canyon (Figures 50; 51). The slide material consisted of Precambrian gneiss, schist, and dolomite with foliation

FIGURE 48 Duck Creek fault scarp formed at the time of the Hebgen earthquake. In the early morning of August 18, 1959, the white Cadillac roared off the edge of the scarp and turned over. The driver, coming from the opposite direction, could not see that the highway had been offset by the fault. Photograph by W.B. Hall.

FIGURE 49 View west at Red Canyon fault scarp at Blarneystone Ranch (sec. 9, T. 12 S., R. 5 E.). The block of ground at left (south) dropped about 12 feet. The culvert in the foreground crosses Grayling Creek. Photograph by I.J. Witkind, U.S.G.S.

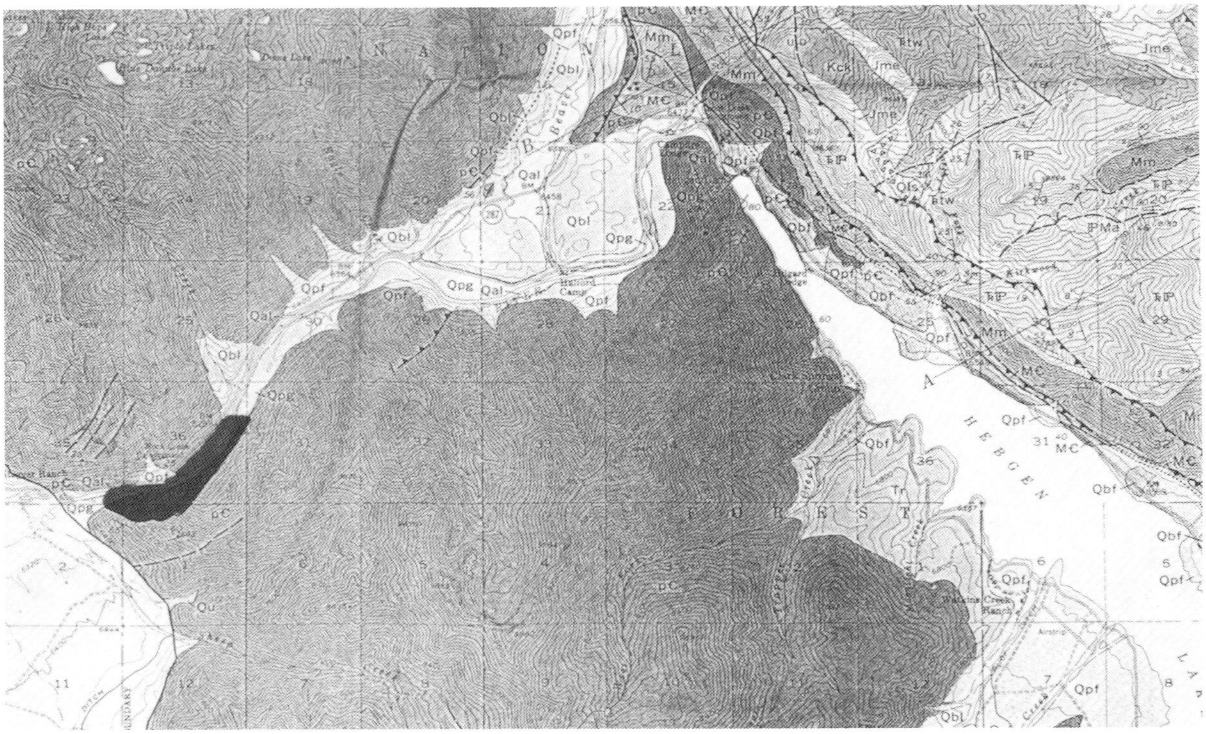

FIGURE 50 Part of the geologic and topographic map of the Hebgen Lake slide area. Black area is the slide. Dark gray areas are Precambrian schist, gneiss, granite, and dolomite. For other symbols see source map (Witkind, and others, 1964, Pl. 5).

and fracture planes parallel to the steep south side of the canyon. Slide debris covered an area of about 130 acres to an average depth of 150 feet. Quake Lake, impounded behind the landslide dam, had briefly a depth of more than 200 feet until the upper part of the dam was removed by the Army Corps of Engineers. The volume of debris in the slide is estimated at 37 million cubic yards (the Lower Gros Ventre Slide that will be seen later in this trip has a volume of about 50 million cubic yards). The momentum of the slide carried debris up on the north side of the canyon to 400 feet above the original river level.

The Hebgen slide is geologically important because it is so well documented. Irving Witkind and his U.S. Geological Survey field party were studying the area at the time of the earthquake and slide and continued to be involved throughout the next five years of study and record-gathering. This event also stimulated an on-going study of the geology in and adjacent to Yellowstone National Park. Another important effect of the earthquake was its influence on the thermal features of Yellowstone National Park. Marler (in U.S.G.S., 1964, p. 185-197) stated:

"The 1959 Hegben Lake earthquake resulted in great changes in the functional behavior of most of the thermal springs along the Firehole and Gibbon Rivers in the Yellowstone National Park. The effects were particularly marked along the Firehole drainage. This is the region of Yellowstone's most famous geysers and hot springs. The initial tremors served as an impetus to bring about eruptive activity in scores of geysers and quiescent springs. Many long-dormant geysers erupted; a few new geysers were born. Some hot springs seemed to be affected adversely, but on the whole there was a noticeable increase in thermal energy that, with the passing of months, gave evidence of persisting." He amplified this statement with data on many specific sites.

The route retraces to Madison Junction and turns south up the Firehole Canyon (Figure 20). The walls on the east side are Nez Perce Creek rhyolite flow (160,000 years old) and on the west, the somewhat younger West Yellowstone rhyolite flow (117,000 years old). From the south end of Firehole Canyon, the road emerges onto a broad, partially treeless area, the Lower Geyser Basin, dotted with many active thermal vents, including several geysers. These have been mapped by Christiansen and Blank (1974).

Major volumes of hot water are emitted in this and the other geyser areas and it is thought that most of it is groundwater that sinks along fractures and faults to a depth of 5,000-10,000 feet, is heated by magma, and then circulates back up to the surface with superheated temperatures.

On the east side of the Lower Geyser Basin, east of the highway, is a prominent rounded hill composed of the Biscuit Basin rhyolite flow (about 542,000 years old). From here to Stop 22 at Old Faithful, the rocks on the east side are rhyolite of the Mallard Lake flow (about

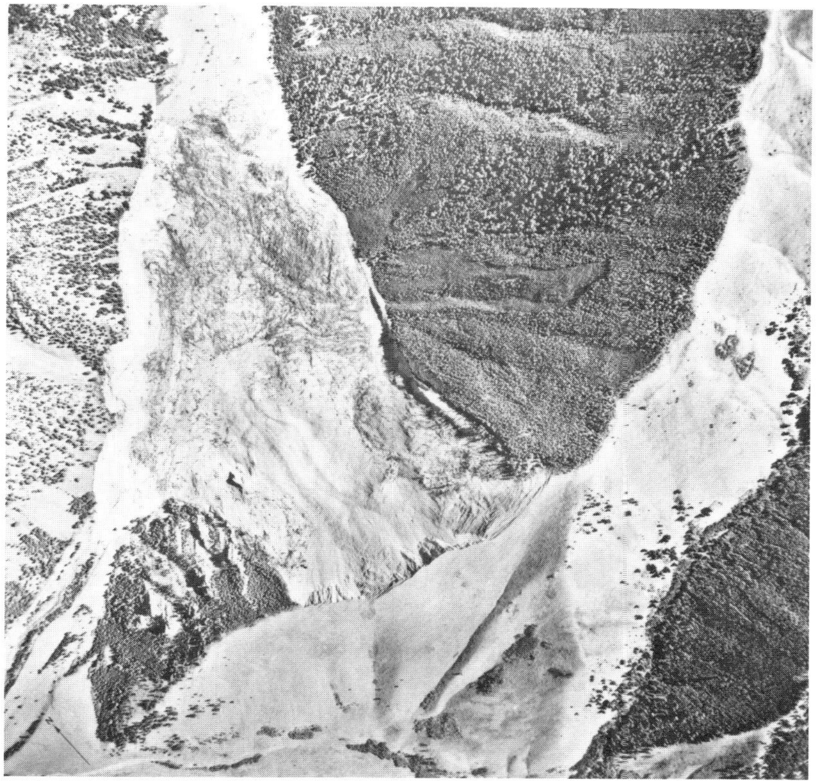

FIGURE 51 Vertical aerial photograph of the Madison Canyon slide. The Recent scarp of the Madison Range normal fault is along the lower left margin. All bedrock is Precambrian schist, gneiss, granite, and dolomite. A section of the near (south) wall of the canyon rim 2,200 feet long and 1,300 feet high slid northward (away from camera) into the canyon and covered an area of about 130 acres to an average depth of 150 feet. U.S Forest Service photograph, Aug. 22, 1959.

150,000 years old) and on the west the West Yellowstone rhyolite flow (117,000 years old). In the Upper Geyser Basin, with Old Faithful geyser near the southeast end, many thermal features emerge from broad areas of siliceous sinter deposits.

Stop 22

The stop is at Old Faithful geyser, the "jewel in the crown" of individual thermal features in the Park. This geyser (Figure 52) has been famous for nearly 120 years because of its dependability in erupting an average of once in every 65 minutes (actually the range is from 33 to 96 minutes). For details on the mechanics of geyser development in general, see Keefer (1972) and for Old Faithful in particular, see Marler (1969). The field trip participants are fortunate to have R.A. Hutchinson, Park Service geologist at Old Faithful, available for detailed discussion and answers to questions.

From Old Faithful the route goes west to West Thumb, first crossing the Scoup Lake rhyolite flow (225,000-275,000 years old), then the Mallard Lake flow to Craig Pass, and finally, the slightly younger Dry Creek flow (160,000-209,000 years old). The last part of the route for Day 5 from West Thumb to Canyon Village is retraced on Day 6 so is not discussed here.

Canyon Village Overnight Stop

The Village is on the Canyon rhylite flow (about 613,000 years old). Figure 53 is a remarkable historic air oblique photograph taken on August 29, 1936, by H.M. Cowling for the Blister Rust Control Division of the Bureau of Plant Entomology, U.S. Department of Agriculture, while he worked for the 116th Photo Section, Washington National Guard. Most of the facilites shown in this photograph have been replaced or modified.

DAY 6

Stop 23

The following discussion includes observations from both the northwest side of the canyon at the overnight stop and also from the southeast side at Stop 23. Table 5 summarizes the very complex Quaternary history of the canyon area (Richmond, 1987). More details are given in

FIGURE 52 Old Faithful geyser erupting during the summer of 1871. Photograph by W.H. Jackson for F.V. Hayden's U.S. Geological and Geographical Survey of the Territories.

his narrative discussion and the regional history related to the canyon-cutting in Richmond (1976).

At the Upper Falls (109 feet – 33 m high) the river flows over a hard ledge in the upper part of the Canyon rhyolite flow about 613,000 years old. Between the Upper and Lower Falls the river has cut through softer strata. At the Lower Falls (319 feet – 97 m high) the river flows over another hard layer in the Canyon rhyolite. From the plunge pool at the base of these falls and farther downstream in the multi-hued main part of the Grand Canyon, the river cuts through much softer, hydrothermally altered rhyolite.

According to Richmond, the time the first phase of canyon-cutting began was about 140,000 to 150,000 years ago. Subsequently, as is shown in Table 5, there were several intervals of sediment deposition at several levels within the canyon. Each was followed by more erosion. These fluctuations were related to periods of ice accumulation and subsequent melting, as well as to tectonic, volcanic, and hydrothermal activity. Details of the regional history as related to canyon-cutting are given by Richmond (1976).

From Stop 23 the route retraces southward to Fishing Bridge along the same highway used on the last of Day 5. The northern part crosses the broad, treeless expanse of Quaternary white, soft, horizontal, unfaulted, tuffaceous fluriatile and lacustrine strata and generally formless glacial deposits in Hayden Valley. The Yellowstone River meanders slowly across the valley, hardly incised at all. This is in sharp contrast to the roaring river in the canyon at Stop 23.

At the southeast margin of Hayden Valley, thermal

FIGURE 53 Air oblique view southwest showing the Upper and Lower Falls of the Yellowstone River and the beginning of the Grand Canyon of the Yellowstone River. Indicated are: Chittenden Bridge across the Yellowstone River, built in 1903 (A), Canyon Village and Lodge as they were in 1936 (B), Lower Falls flowing over a hard ledge in the Canyon rhyolite about 613,000 years old (C), sediments of Inspiration Point dated at about 103,000 years (D), and Upper Falls flowing over a hard ledge in the upper part of the Canyon rhyolite flow (E). Photograph by H.M. Cowling for Washington National Guard.

areas become abundant and an intricate system of northeast-trending and northwest-trending normal faults cut across all the Quaternary rhyolite flows and even some alluvial deposits (Figure 54-A). This is the area of the currently rising crustal dome just north of Fishing Bridge (FB on Figure 54; Christiansen and Blank, 1975). Smith and Braile (1984) measured an uplift rate on this dome of 15 mm per year. Overlapping this area of uplift and extending to the north-northeast for several miles is a seismic low velocity zone (Figure 55), concerning which they state (1984, p. 107): "The pocket of low-velocity, ~ 4.0 km/sec material beneath the northeast caldera boundary suggests that a fluid-filled body of steam-water, partial melt, or both has existed recently. Whether it is in a stage prior to an eruption cannot be determined from the available data. However, the trend of increasing crustal uplift (at rates greater than 15 mm/yr and in excess of normal nonthermal processes) toward the northeast points toward this low-velocity body. The coincidence of the low-density, low-velocity body near an area of increasing uplift suggests that the northeast caldera could be a candidate for eruption."

Stop 24

This stop is along the emerging north shore of Yellowstone Lake which is one of the largest natural mountain lakes in the United States, with a shoreline 110 miles long and a surface area of about 137 square miles. Its depth is generally less than 200 feet except east of Stevenson Island where it exceeds 300 feet. The present and past shorelines and sedimentary deposits in and marginal to the lake, which is surrounded by an active or recently active thermal, volcanic, tectonic, and glaciated area, record an incredibly complex history of Quaternary events. Researchers for the last 100 years and more have been trying to assemble enough data to unravel the history of the area. Just since the 1950s, air photos, new topographic maps, regional and detailed geologic maps, radiometric dating techniques, sophisticated chemical and spectrographic analyses, limnological studies, geophysical and geothermal studies, all have provided a wealth of new data. Still, like blind men feeling different parts of the elephant, even though there is agreement on broad concepts, there is widespread disagreement as to

TABLE 5 Stratigraphy and erosional history of the Grand Canyon of the Yellowstone River from Chittenden Bridge on the south to Sevenmile Hole on the north. Slightly modified from Richmond, 1987, p. 156.

TIME DIVISIONS		UPPER SEGMENT	MIDDLE SEGMENT		LOWER SEGMENT	σO^{18} STAGES
			Late glacial and post-glacial flood, landslide, lake, mudflow, and talus deposits			1
Wisconsin	Late	Pinedale erratics	pkl- Pinedale ice-contact lake silt pt- Pinedale Till (Deckard Flats advance (<~15>) Earlier Pinedale glaciation			2
Wisconsin	Middle		~~ Erosion of Canyon ~~	C	Erosion of Canyon	3
Wisconsin	Early					4
"Eowisconsin"		bl-Varved silt	Sediments of Uncle Toms Rest Area	W C		5a
"Eowisconsin"		Glacial rubble	br- Glacial rubble			5b
"Eowisconsin"		Rh-Hayden Valley rhyolite flow [102:3] Lake silt (11) C W	uf-Sediments of Upper Falls: bslu-Upper sand [103:4] bll- Upper middle silt bisl- Lower middle sand bll- Lower middle silt	W C C	W gv- Sediments of Grand View Erosion of Canyon	5c
"Eowisconsin"			bgl- Middle sandy gravel (erratics)			5d
Sangamon			s-bl-Lower silt	W C	W rr- Sediments of Red Rock ~~ Erosion of paleocanyon ~~	5e
Illinoian	Late		ssg- Lower sandy gravel		Caldera lake rises and overflows to north Rhyolite flows build up Central Plateau; block caldera lake outlets	6
Illinoian			Pumice uf-III [286:13]		Ip- Sediments of Inspiration Point	7
Illinoian	Early		ufd- Diamicton			8
Pre-Illinoian			Pumice uf-II Pumice uf-I			7
Pre-Illinoian			cc-Sediments of Cascade Creek Rc- Tuff of Sulphur Creek; Canyon rhyolite flow lf- Sediments of Lower Falls: Diamicton; Lake sediments Rut- Tuff of Uncle Toms Trail Sediment inclusions in tuff of Uncle Toms Trail	W W W		15
			Lava Creek Tuff [610 ka] and Yellowstone Caldera			

EXPLANATION
C Pollen records parkland at treeline or tundra
W Pollen records parkland below treeline or closed forest
~~ Interval of erosion
<~15> estimated age x 10^3
[103:4] K-Ar age x 10^3

Letter symbols same as U.S.G.S. Map I-652

timing, extent, causes of major events, and correlation with those in other areas. These disagreements include the detailed history of the lake, itself, various glaciations and their relation to tectonism, etc.

The narrative for this field trip is not the forum for evaluating the validity of these interpretations but its purpose is only to call attention to their existence and to encourage further research. Ongoing observations indicate that the floor of Yellowstone Lake has been and still is tilting. Trees, still rooted, are being submerged by sinking of the lake floor along the southeastern margin of the lake, just in the last 20 years. Conversely, with a few local exceptions (Meyer, 1986), the northern part has been rising (Figure 56). Logs buried in near-shore lacustrine strata (Figure 57), presumably deposited in a northern extension of Yellowstone Lake four miles northeast of Mary Bay are at an elevation of 7,840 feet (140 feet above the present lake level) and have a ^{14}C age of 7,550 ± 350 years (Meyer Rubin, written commun., 1965, sample W-1684). Supporting ages come from adjacent localities. If this rise is averaged, it would be one foot in 54 years.

Hydrothermal-explosion craters are among the most spectacular late-glacial and post-glacial events that have occurred in central Yellowstone National Park. At least 10 are known, ranging in diameter from a few hundred feet to perhaps several miles (Muffler, and others, 1971; Wold, and others, 1977). Some of the explosions are thought to have been triggered by an abrupt decrease in confining water pressure by sudden draining of the overlying lake. Several, however, do not seem to fit this interpretation. The one at Squaw Lake (now called Indian Pond) is younger than 13,650 ± 600 years (Meyer Rubin, written commun. to G.M. Richmond, 1973, sample W-2894) and older than 3,500 ± 250 years (Meyer Rubin, written commun. to G.M. Richmond, 1972, sample W-2734). This nearly circular crater is 1,300 feet in diameter and angular fragments of explosion debris can be seen on the inner slopes. Hot water seeps (128°F, 54°C) occur on the east side near lake level (Muffler, and others, 1971, p. 732).

The Turbid Lake hydrothermal-explosion crater, studied in the early 1960s (J.D. Love, manuscript report used by G.M. Richmond, 1976), about three miles east of Squaw Lake (Figures 58; 59) has been dated as 8,310 ± 300 years (Meyer Rubin, written commun. to J.D. Love, 1966, sample W-1944). On the east side, the crater rim rises more than 200 feet above the lake. No overlying lake deposits have been identified. By far the most graphic description of Turbid Lake and vicinity, including a map, is by Comstock (*in* Jones, 1875, p. 190-191). He mentioned that the temperature of the lake was, at that time, 180° to 192°F (82° to 89°C). Muffler, and others, (1971, p. 732) noted that acidic hot springs are on the southeast side and that the lake itself is acidic. This is especially interesting because Comstock (p. 92) reported that on Sedge Creek (he called it Wolf Run Creek) a few hundred yards north of Turbid Lake "there are here a number of cold springs containing sodium chloride, or common salt". It would be interesting to verify this, as salt is rare in Yellowstone National Park except in Cretaceous strata near the south border, where there are

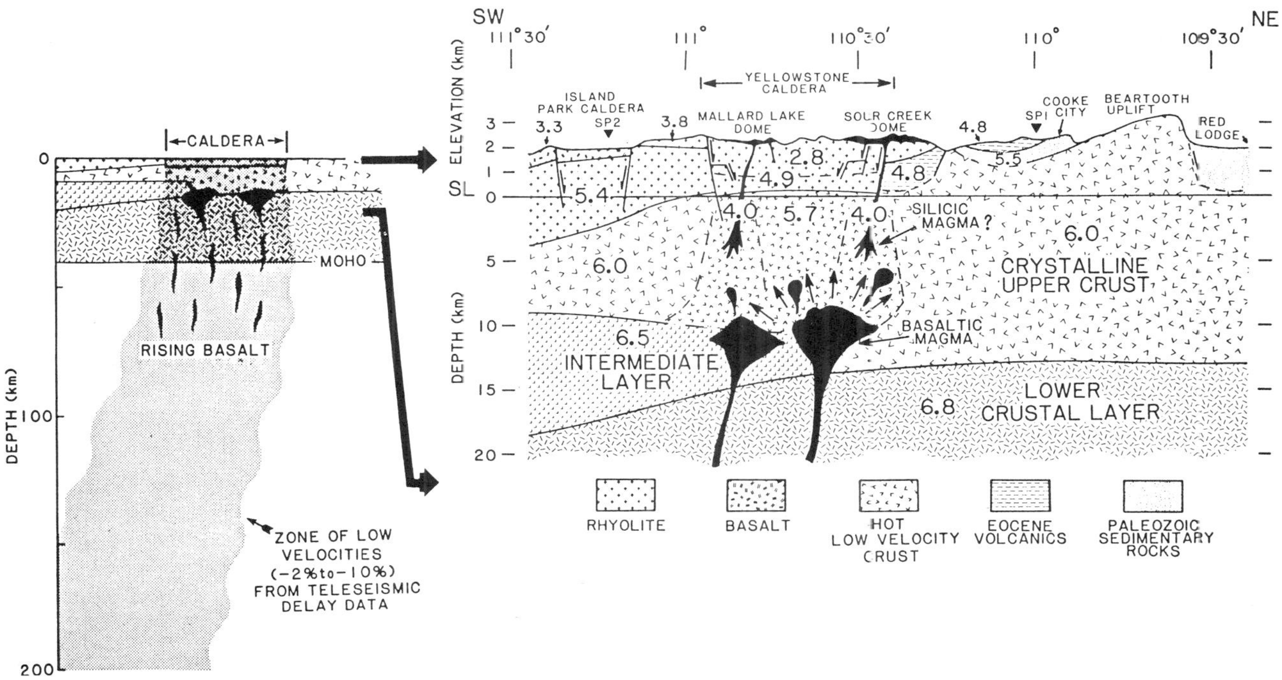

FIGURE 54-A Idealized northeast-southwest geologic-seismic velocity model for the crustal structure of the Yellowstone-Island Park-Snake River Plain region. P-wave velocities are in kilometers/second. Geologic interpretations are based on constraints of the Quaternary history and the petrologic models of Leeman (in press) and Christiansen (U.S. Geological Survey, personal communication) After Smith and Braile, 1984.

a number of "elk licks". No Cretaceous rocks are known in the central part of the Park.

The relation of the hydrothermal-explosion deposits around Turbid Lake and their age with respect to what is known of the fluctuations of Yellowstone Lake level, make it seem unlikely that this explosion was caused by lowering of this part of the lake 8,300 years ago.

Mary Bay is considered by some workers (Wold, and others, 1977) to be the northern part of a large hydrothermal explosion crater. This is based on bathymetric and geophysical evidence. In the 1980s, several groups of geochemists and microbiologists from the University of Wisconsin at Milwaukee, and from other universities have explored a series of fissures on the lake bottom in the deepest part of the lake (about 365 feet), east of Stevenson Island. Some are reported to have sheer walls more than 100 feet high. One orifice at a depth of 320 feet spouted water with a temperature of 208°F (98°C). Many unique organisms are present near these deep-water thermal features (National Park Service, press release, 1988).

A short optional side trip via paved road climbs about 500 feet from the broad open treeless valley at Mary Bay to the top of Lake Butte (elevation 8,348 feet). This point provides a view of Yellowstone Lake and the regional geology across the lake in all directions. The butte, itself, is capped by hard gray andesitic volcaniclastic rocks, the Langford Formation of middle Eocene age (about 48 million years old). This sequence has been intruded by a fine-grained gray rhyodacite porphyry, likewise of Eocene age.

On the way to Lake Butte the route passes Steamboat Point and Springs. The rock in the highway cut at the turnout is a hodgepodge of large and small angular blocks of older lithified debris of uncertain age and correlation, blown out in a tremendous hydrothermal-explosion (probably from the Mary Bay crater) from depths as yet unknown. Regarding this thermal feature, Comstock (1875, p. 192) states: "The trail runs above several more springs, one of which is the Steamboat, appropriately named from the peculiar sound emitted, which closely resembles the puffing produced by the escape of steam from the cylinders of a high pressure steamer. This is caused by the passage of the vapor from a subterranean

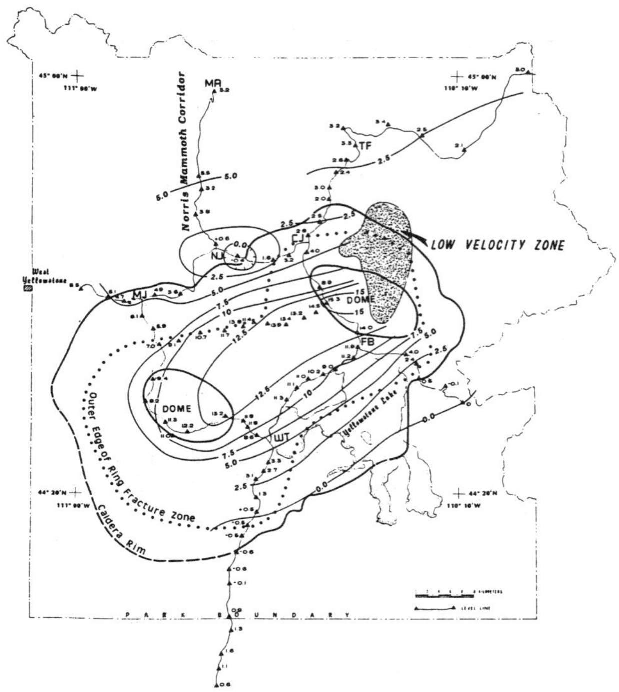

FIGURE 54 Contour map of relative uplift rates over the Yellowstone Plateau. Contours are in millimeters/year and were computed from the data of Smith and Pelton (1979, 1982). Bench-mark locations shown by triangles; MA, Mammoth; NJ, Norris Junction; C.J., Canyon Junction; FB, Fishing Bridge; WT, West Thumb; TF, Tower Falls. From Smith and Braile, 1984.

chamber, under pressure, through a narrow orifice in the hard rock. A few rods farther up the lake, sulphurous vapor issues from a cavern, in conglomerate, with a seething noise, as it comes in contact with the waves, not unlike the sound produced by the escape of steam from the cylinders of a locomotive moving slowly. For the purposes of distinction, this will here be called the Locomotive Spring."

From Stop 24 the route retraces to West Thumb and then turns south into the area covered in Figure 60. Most of the route to West Thumb is on dark gray obsidian sand and white lake sediments, both of Quaternary age. At Bluff Point, however, for a mile the highway crosses an unnamed ash flow tuff with an age of 162,000 years (Christiansen, written commun.). Duck Lake, on the northwest side of West Thumb village is a small oval-shaped hydrothermal-explosion crater surrounded by explosion debris. West Thumb arm of Yellowstone Lake is considered to be a large caldera (Christiansen, 1984). Farther south the route crosses the Continental Divide on formless glacial deposits at an elevation of 7,988 feet, from drainage going eastward to the Mississippi River to that going westward into the Pacific Ocean. All along the east shore of Lewis Lake are exposures of the Aster Creek rhyolite flow (about 174,000 years old). Lewis Falls drops over a hard ledge in this flow. The route here is approximately along the buried trace of the southern margin of the Yellowstone caldera, and is just west of the Red Mountains caldera (Christiansen, 1984). Still farther south, the prominent dark gray cliffs along the west side of the highway are part of the Pitchstone Plateau rhyolite flow, the youngest (70,000 years) of the major flows in Yellowstone National Park.

The Lewis River, a tranquil stream east and just below the highway level, suddenly becomes a raging torrent southward, descending 800 feet in five miles and cutting the Lewis River Canyon as much as 500 feet deep through hard rhyolites. The upstream part of the canyon cuts the Lava Creek Tuff composed of rhyolite ash flows about 615 million years old. The lower, downstream part of the canyon, cuts through the Lewis Canyon Rhyolite flow (929,000 years old; Christiansen, written commun.). Near the south entrance of the Park, at the junction of the Lewis and Snake Rivers, there is a concealed normal fault that dropped these volcanic sequences down against Paleozoic strata that are part of the Basin Creek uplift (Love and Keefer, 1969), a structure which is separate from any of the adjacent mountain ranges.

From the south boundary of Yellowstone National Park the route follows a transition zone into Jackson Hole. Beginning about four miles south of the Park boundary, the highway is flanked by the Pliocene Huckleberry Ridge Tuff (2.1 million years old), an ash flow tuff that rests directly on Cretaceous rocks here. This tuff has regional continuity and is the most useful sequence in the region for working out the timing of Quaternary tectonic events. To the west across the north end of Jackson Lake, the north end of the Teton Range becomes prominent, rising southward, and diving under the Pitchstone Rhyolite plateau northward into Yellowstone National Park. At the base of the Tetons, the break in slope marks the Teton normal fault whose movement began less than nine million years ago; it is still moving. The Huckleberry Ridge Tuff dips west here into Jackson Lake (surface elevation 6,769 feet), yet at Berry Peak (elevation 8,951 feet), four miles to the west, the tuff likewise dips west. In between is the Teton normal fault complex. Regional structure contour data on top of the tuff suggest about 5,000 feet of displacement along this part of the Teton fault (Teton block up, Jackson Hole block down) in the last two million years. This would be an average of one foot of displacement every 400 years.

The view west across the north end of Jackson Lake shows Paleozoic and Mesozoic rocks wrapping around the north end of the Teton Range and diving under Jackson Lake. Where the gray cliffs of Madison Limestone intersect the lake, and are cut off there by the Teton fault, a half-mile-long series of hot springs emerges from the fault trace. The average temperature of 27 orifices is 135°F (57°C), with a maximum of 156°F (69°C). The estimated combined flow of the springs above the lowest water level is about 2.4 million gallons per day but many more orifices can be seen out in the lake so the total flow is actually much larger. Four of many gas seeps emerging

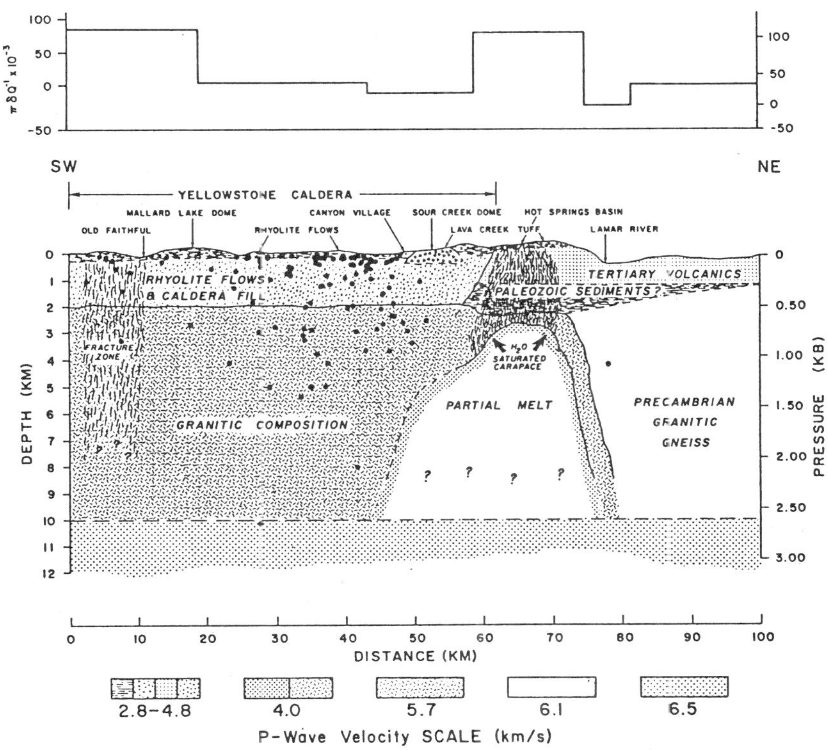

FIGURE 55 Generalized SW-NE upper-crustal cross section parallel to the axis of the Yellowstone caldera. Focal depths (shown by black dots) were projected from outside the low-velocity body. Relative P-wave seismic attenuations are averaged from data of Clawson and Smith (1981). From Smith and Braile, 1984.

FIGURE 56 Emerging shoreline along the north margin of Yellowstone Lake. Air oblique view northwest toward Fishing Bridge and Yellowstone River outlet to the lake. Beyond Fishing Bridge, several terrace levels are visible in the treeless area. These have been described and measured by Meyer (1986). Photograph by National Park Service.

FIGURE 57 Twelve feet of near-shore lacustrine brown, gray, and indigo blue clay, silt, and sand with a 1-foot bed of gravel in middle (at upper wood horizon). A tree trunk one foot in diameter in this bed yielded a ^{14}C age of 7,550 ± 350 years (Meyer Rubin, written commun., 1965, sample W-1684). Site is on east bank of Astringent Creek, 500 feet upstream (north) from old service road crossing. Photograph by J.D. Love, July 16, 1965.

FIGURE 58 View northeast across Bear Creek at escarpment of hydrothermal-explosion deposits one-fourth mile east-southeast of Turbid Lake. Indicated are: 85 feet of conglomerate and rusty brown sand (B), 20 feet of blue-green sand (C), till (?) or explosion breccia (D), two inches of charcoal with ^{14}C age of 8,310 ± 300 years (Meyer Rubin, written commun., 1966, sample W-1944; later sampling of this same site by Richmond (1976, p. 31) confirmed this age), site of closeup view in Figure 59 (E), and explosion breccia 10 feet thick (F). Photograph by J.D. Love, October 4, 1966.

FIGURE 59 Closeup view of site E in Figure 58, showing two-inch black charcoal bed where sample for ^{14}C age was collected. The carbon zone is overlain by hydrothermal-explosion deposits and underlain by till (?) or explosion breccia. A sight reading along 100 feet of outcrop of the carbon layer shows Str. N. 80° E., Dip 15° N. Photograph by J.D. Love, September 19, 1967.

with the hot water were sampled. Each contains a distinctive nitrogen and carbon ratio, suggesting complex origins in different parts of this hot spring system (Kvenvolden, and others, 1989).

All the stops in the Jackson Hole area except for 31 and 32 (Figure 60) are within the boundary of the Grand Teton National Park geologic map (Love, and others, in press). This map probably will not be printed in time for Field Trip T-328 but a preliminary copy will be available for discussion. Geologic maps of many individual quadrangles are, however, available, as is a booklet on the origin of the Teton landscape (Love and Reed, 1972); road logs of Jackson Hole and the Gros Ventre River Valley (Love, and Love, 1983; 1988), and a geologic block diagram of the Teton-Jackson Hole region with events tied in to a geologic column (Love, and others, 1973).

The Teton mountain front is the youngest (less than 9 million years), and steepest of the high mountain fronts in the Rocky Mountain chain. The unique scenery of the Tetons and adjacent areas is the product of an unusual geologic history. The record of sedimentation, volcanism, and tectonism, especially that beginning with Late Cretaceous time, shows one of the thickest (more than 50,000 feet) most complete, and most complex sequences in North America. Tectonism did not end with the Laramide orogeny in Eocene time but has continued intermittently to the present. This, in part, explains the fresh jagged appearance of the Teton Range, the anomalous drainages, the abundant landslides, the narrow canyons, and other distinctive features.

Figure 61 shows the rugged nature and asymmetry of the Teton Range, the ice-carved canyons, and the fresh, unmodified rock glaciers. Mount Moran (Figure 62; elevation 12,605 feet) is composed of dark migmatitic biotite gneiss (>2,875 million years). It is cut by many light-colored dikes of Mount Owen Quartz Monzonite (about 2,500 million years). The vertical black diabase dike, about 150 feet wide, is 775-1,350 million years old. This is the youngest Precambrian rock in the Teton Range,

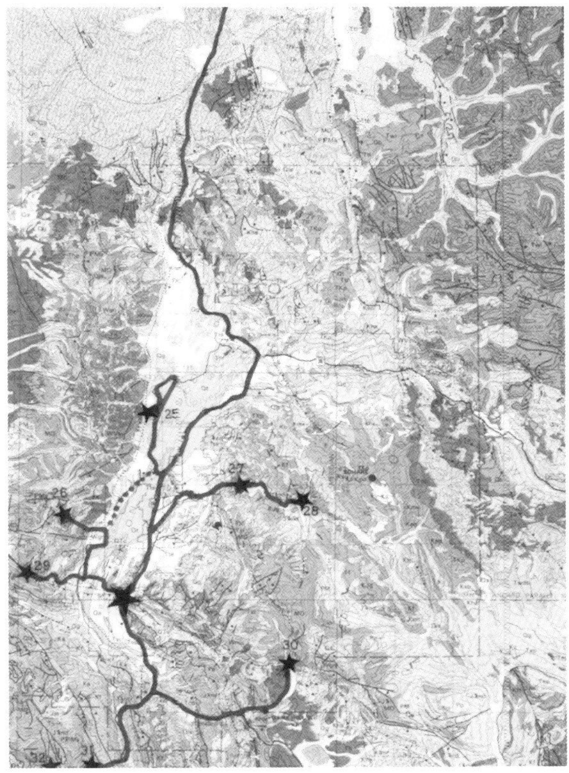

FIGURE 60 Segment of the geologic map of Wyoming (Love and Christiansen, 1985) showing the route of the field trip (heavy black line) on most of Day 6, all of Days 7 and 8, and the first part of Day 9. Large star indicates overnight stop at Jackson, Wyoming. Smaller numbered stars show stops along route. For geologic symbols and line patterns see source map. Squares are approximately six miles on a side.

FIGURE 61 Oblique aerial view south along the Teton Range showing the steep east face and the gentler west side. Rocks in most of the panorama are Precambrian metasedimentary and metavolcanic, older than 2,875 million years. Grand Teton (elevation 13,770 feet) is the high peak at the top center. West-dipping Palozoic rocks are at top right. Jackson Lake in Jackson Hole is at upper left. Prominent rock glacier in foreground may be underlain in part by ice. All of the terrain has been extensively glaciated. Photograph by J.S. Shelton.

and has been traced westward for more than seven miles. Partly overlapping the black dike is a gray mound consisting of about 50 feet of nearly horizontal Flathead Sandstone of Middle Cambrian age (Figure 63). This occurrence is especially significant from a structural standpoint because it provides the highest control point on the Cambrian-Precambrian surface in the northern part of the Teton Range and helps to determine the general amount of displacement on the Teton fault system. East of a line between Mount Moran and the Grand Teton, the total displacement is thought to be between 30,000 and 35,000 feet. Most or all of this movement probably occurred during the last nine million years. Falling Ice glacier is at the left, partly covering the black dike. Skillet glacier is at the right. Figure 63 shows, on top of Mount Moran, a small flat remnant of the Precambrian peneplain on which the Middle Cambrian sea advanced and deposited the Flathead Sandstone 550 million years ago.

From Moran Junction south to Jackson, the highway is chiefly on outwash quartzite gravel deposited by meltwater from several glaciations. Details of the geology along this route are in Love, and Love (1983).

Jackson, Wyoming Overnight Stop

Because the stops during the next two days radiate out in all directions from this overnight stop, a summary of Upper Cretaceous and Cenozoic stratigraphy, tectonic, and glacial events, from oldest to youngest, in Jackson Hole, is presented here (abbreviated from Love, 1977). The discussions during Days 7 and 8 are keyed in to these numbered depositional and tectonic events.

Depositional Event 1. FRONTIER FORMATION (UPPER CRETACEOUS). Gray to black partly marine sandstone and shale, 1,000 feet thick.

Depositional Event 2. CODY SHALE (UPPER CRETACEOUS). Gray to black marine shale, 1,300-2,200 feet thick.

Tectonic Event 1. In eastern Idaho, the Targhee uplift

FIGURE 62 Oblique aerial view northwest at east face of Mount Moran. Vertical relief from bottom of photograph to summit at 12,605 feet is about 3,400 feet. Most of the rock is migmatitic biotite gneiss more than 2,875 million years old The vertical black diabase dike, 150 feet wide, has a controversial age, from 775 to 1,350 million years. The Flathead Sandstone of Middle Cambrian age, is the nearly horizontal gray mound that partly overlaps the black dike (see Figure 63 for close-up view). Falling Ice glacier is at left, partly covering the black dike; Skillet glacier is at right. Photograph by Austin S. Post, University of Washington, Aug. 19, 1963.

rose and was eroded to its Precambrian core in middle Niobrara time. Gold-bearing quartzite debris spread eastward for the first of many times into Jackson Hole and intertongued with marine and brackish-water strata.

Depositional Event 3. BACON RIDGE SANDSTONE (UPPER CRETACEOUS). Light gray marine and brackish water sandstone, gray shale, coal. Locally has gold-bearing quartzite clast conglomerates. Thickness 900-1,200 feet.

Depositional Event 4. SOHARE FORMATION (UPPER CRETACEOUS). Nonmarine lenticular gray sandstones and shales; abundant coal in lower part. Thickness about 3,500 feet.

Tectonic Event 2. Recurrent movement on the Targhee uplift caused several sequences of gold-bearing quartzite-clast conglomerates to spread eastward into Jackson Hole.

Depositional Event 5. MESAVERDE FORMATION (UPPER CRETACOUS). White massive soft nonmarine sandstone with thin shale and coal beds. Several gold-bearing quartzite conglomerates. Thickness 0-1,000 feet.

Tectonic Event 3. Recurrent movement on the Targhee uplift caused several sequences of gold-bearing quartzite conglomerates to spread eastward into Jackson Hole. Somewhere to the west volcanos erupted large quantities of ash that blew into Jackson Hole.

Depositional Event 6. MEETEETSE FORMATION (UPPER CRETACEOUS). Chalky white tuffaceous sandstone, tuff, and yellow bentonite beds; several gold-bearing quartzite conglomerates. A K-Ar age of one tuff bed is 73.1 million years.

Tectonic Event 4. The ancestral Teton-Gros Ventre uplift rose as a broad northwest-trending arch connecting the sites of the present Gros Ventre and Teton Ranges with the much higher, broader, and older Targhee uplift to the northwest. The Basin Creek uplift along the southern border of Yellowstone National Park northeast

FIGURE 63 Close-up aerial view southwest of the top of Mount Moran showing the overlap of the Middle Cambrian Flathead Sandstone across part of the black dike. The flat surface is a small remnant of the Precambrian peneplain across which the Cambrian sea advanced 550 million years ago. West-dipping Paleozoic rocks on the west flank of the Teton Range are at top, and Teton Basin is in the distance. Photograph by J.S. Shelton.

of the ancestral Teton Range rose enough to cause stripping of more than 5,000 feet of Cretaceous rocks before deposition of the younger Cretaceous sequence.

Depositional Event 7. HAREBELL FORMATION (UPPER CRETACEOUS). Gold-bearing quartzite conglomerates from the Targhee uplift interbedded with tuffaceous sandstones and shales. Some marine and brackish-water fossils in finer-grained beds. Thickness 0-10,000 feet.

Tectonic Event 5. The Harebell depositional trough sank so much that by the time the youngest strata in the formation was deposited, the oldest were, in places, 10,000 feet below sea level. Volcanos in or north of Yellowstone National Park supplied debris to Jackson Hole. The Washakie Range rose as a northwest-trending arch along the northeast margin of Jackson Hole, was thrust westward, and then eroded to its Paleozoic core, all within latest Cretaceous time. Probably simultaneously several steep-sided northwest-trending anticlines in eastern Jackson Hole were formed, thrust westward, and stripped of 10,000-15,000 feet of Cretaceous rocks before deposition of Paleocene strata. The Basin Creek uplift rose as much as 12,000-15,000 feet and all the Harebell Formation and 6,000 feet of older Mesozoic rocks were eroded off the crest. A major volcanic eruption occurred somewhere upwind and supplied ash to the basal part of the Pinyon Conglomerate.

Depositional Event 8. PINYON CONGLOMERATE (UPPERMOST CRETACEOUS AND PALEOCENE). Gold-bearing brown quartzite conglomerate with some sandstone and claystone. A biotite-rich tuff 300 feet above the base has a K-Ar age of 67 million years. A Cretaceous dinosaur tooth, however, was found 150 feet above the base. In southeastern Jackson Hole, fossils in the basal beds show that the entire section there is Paleocene. Thickness is 0-3,700 feet.

Depositional Event 9. DEVILS BASIN FORMATION (PALEOCENE). Sandstone and claystone, greenish-gray and brown, variegated in places at top; intertongues at base with Pinyon Conglomerate. It is a partial time equivalent of the Hoback Formation on the south side of the Gros Ventre Range. Thickness is 1,000-2,000 feet.

Tectonic Event 6. Recurrent uplift of the Gros Ventre Range resulted in development of an asymmetric mountain arch, steep on the southwest flank. The Skyline Trail

Conglomerate Member of Dorr and others, (1977) was deposited as a torrential fan. This depositional event occurred concurrently with the tectonic event so is not listed separately.

Tectonic Event 7. The Jackson thrust plate moved 50 miles, more or less, northeastward and overrode the Skyline Trail Conglomerate Member. At the same time or perhaps slightly earlier, the Little Granite anticline was formed and then was overridden by the Jackson thrust plate. The timing of this thrusting had to be earliest Eocene because the next younger tectonic event had at least begun before early Eocene time.

Tectonic Event 8. The rise of the Gros Ventre uplift that caused deposition of the Skyline Trail Conglomerate culminated in southwestward thrusting of the Gros Ventre Range along the Cache Creek thrust.

Despositional Event 10. UPPERMOST PALEOCENE STRATA AND LOWER EOCENE INDIAN MEADOWS AND WIND RIVER FORMATIONS AND EQUIVALENTS. Three sequences are present: lower variegated sequence of red and white claystones and thin sandstones, 1,200 feet thick containing latest Paleocene and earliest Eocene mammals; a middle coal, dark gray shale, sandstone, and fossiliferous limestone sequence 100-300 feet thick; and an upper variegated sequence of slightly tuffaceous, red, purple, and white claystone, sandstone, and conglomerate of Paleozoic clasts, 800-1,100 feet thick, of early Eocene age.

Tectonic Event 9. During deposition of part of the Wind River Formation (Lysite age), the Targhee uplift rose further and an enormous flood of gold-bearing quartzite clasts was carried by a large river system across the southeast part of Jackson Hole, along the northeast side of the Gros Ventre Range, and around the southeast end of the range. As much as 3,500 feet of this conglomerate now composing the Pass Peak Formation, was deposited in the rapidly sinking northern part of the Green River Basin. At the same time the adjacent part of the Gros Ventre Range rose and large angular blocks of Paleozoic and Mesozoic rocks from the uplift were incorporated in the conglomerate. Strangely, the strata that are the time equivalent of these rocks a few miles to the east do not reflect this sedimentation.

Tectonic Event 10. Possibly the New Fork thrust fault extended northwest from the Wind River Range at this time and may have thrust the southeast end of the Gros Ventre Range on top of the Pass Peak Formation (Mills, 1989, Fig. 5).

The following tectonic and depositional events do not include those affecting the eastern part of Jackson Hole. To include them would make this summary too long.

Depositional Event 11. HOMINY PEAK FORMATION (MIDDLE EOCENE). As much as 2,000 feet of mafic volcaniclastic and igneous rocks, deposited at the north end of the Teton Range and near Moran Junction in the eastern part of Grand Teton National Park.

Tectonic Event 11. The Targhee uplift in eastern Idaho, after repeatedly providing hundreds of cubic miles of gold-bearing quartzite conglomerate debris to Jackson Hole for 35 million years (from 83 to 48 m.y.), subsided and never again provided debris to Jackson Hole. The amount of subsidence was about 6,000-8,000 feet. As a result, the east flank of the ancestral Teton Range in middle Eocene time became the west flank, much as it is seen today.

Depositional Event 12. WHITE RIVER FORMATION (OLIGOCENE). This formation is known from only two small outcrops, one on the northeastern margin of Jackson Hole and the other in the north-central part where less than 100 feet of white tuff is present. Its K-Ar age is 35.8 million years.

Depositional Event 13. COLTER FORMATION (LOWER AND MIDDLE MIOCENE). More than 5,000 feet of volcaniclastic rocks, mostly mafic, derived chiefly from a now-buried major vent somewhere near the northern margin of Jackson Lake (Barnosky, 1984).

Tectonic Event 12. During deposition of the Colter Formation, that part of the present Jackson Hole area lying east and southeast of Jackson Lake sagged several thousand feet along a north-trending trough line. This was the first major downwarping of the area after Paleocene time.

Tectonic Event 13. At about the beginning or during the early part of late Miocene time, the southwest margin of the Gros Ventre Range and adjacent highlands in the Hoback Range were uplifted and shed coarse debris southwestward to form conglomerate in the Camp Davis Formation. Masses of Mesozoic rocks several miles in extent broke loose from these uplifted areas, moved southwestward across the conglomerates, and were buried by more conglomeratic debris. No general agreement exists as to whether these Mesozoic rock masses are klippen from a reactivited Cache Creek thrust plate or whether they are detachment plates that moved by gravity sliding. If the latter, there are problems in identifying the source areas.

Despositional Event 14. CAMP DAVIS FORMATION (LOWER (?) UPPER MIOCENE). The Camp Davis Formation consists of more than 5,500 feet of red conglomerate, with a basal gray conglomerate overlain by white limestone, white vitric rhyolite tuff, and fossiliferous gray claystone. The thrust or detachment masses of Mesozoic rocks described in Tectonic Event 13 were emplaced in the upper conglomerates.

Tectonic Event 14. The breakup of the ancestral Teton-Gros Ventre uplift began with movement along the Warm Spring fault north of the Gros Ventre Buttes about five miles north of the town of Jackson. The part of Jackson Hole north of the fault continued to sag, eventually as much as 6,000 feet, and became the site of Teewinot Lake, in which the Teewinot Formation accumulated.

Depositional Event 15. TEEWINOT FORMATION (UPPER MIOCENE). White limestone, white vitric rhyolite tuff, and gray claystone, fossiliferous, 0-6,000 feet thick, with K-Ar age of nine million years in the upper beds. Fine-grained strata in the upper part are only two miles from the present steep east flank of the Teton Range as it now exists. If it had existed during Teewinot deposition, coarse debris would have been deposited this close to the mountain front. Because there is none, it is assumed that the mountain front is less than nine million years old. The Teewinot Lake Basin continued to sink north of and against the Warm Spring fault and by the time the youngest Teewinot strata were deposited, the oldest were below sea level.

Tectonic Event 15. Teewinot Lake was destroyed by foundering of the middle part of the ancestral Teton-Gros Ventre uplift along a series of north-trending normal faults. These displace the slightly older east-trending Warm Spring fault. The Teton fault system developed at a 55° angle to the axis of the ancestral Teton-Gros Ventre uplift, and extended across it approximately along the eastern base of the present Teton Range. Local displacement was 30,000 to 35,000 feet, with the Jackson Hole block hinging down and tilting westward and the Teton block rising and likewise tilting westward. Differential movement along subsidiary faults resulted in development of a series of west-tilted fault blocks of resistant Paleozoic rocks such as the Gros Ventre and Blacktail Buttes.

Local igneous intrusions of several types are along and near faults. For example, obsidian in two localities in southern Jackson Hole have K-Ar ages of eight and nine million years, respectively, and an andesite intrusive along the highway between Wilson and Teton Pass has an age of eight million years. Several igneous masses on East and West Gros Ventre Buttes have not been dated but have been displaced by faults that collapsed the ancestral Teton-Gros Ventre uplift.

Depositional Event 16. RED CONGLOMERATE (MIOCENE OR EOCENE). On high surfaces south of the Gros Ventre Range, possibly as much as 2,000 feet thick. No fossils have been found, but the sequence looks more like the upper red part of the Miocene Camp Davis Formation and less like the Eocene Skyline Trail Conglomerate Member of the Hoback Formation. No tuffaceous beds have been reported. It is folded and faulted in places but its relation to thrusts is controversial. The faulting that hinged down the Camp Davis Formation along the Hoback normal fault, with more than 5,000 feet of displacement, may have occurred at this time.

Depositional Event 17. CONANT CREEK TUFF (UPPER MIOCENE OR LOWER PLIOCENE). Ash flow rhyolite tuff representing the first of several major rhyolite eruptions in Jackson Hole and Yellowstone National Park. Its age has been reported as 4.2 to 5.89 million years, hence either latest Miocene or early Pliocene.

Depositional Event 18. SHOOTING IRON FORMATION (PLIOCENE). Lacustrine and fluviatile brown, red, and gray highly tuffaceous claystone, sandstone, and thin conglomerates; found for certain only in southern Jackson Hole and on the Gros Ventre Range.

Tectonic Event 16. After depositional of the Shooting Iron Formation, vertical movement of about 4,000 feet took place. Much of it probably was down-sagging of Jackson Hole along the Teton and other faults, and some of it may have been uplift of the western end of the Gros Ventre Range, where a remnant of the formation has been identified at 10,000 feet. On the floor of Jackson Hole it is now at 6,000 feet or lower.

Glaciation Event 1 (PLIOCENE). Till of the informally named "Ghost Glaciation" on the floor of Jackson Hole at the north end of Signal Mountain, underlying the Huckleberry Ridge Tuff. Remnants of it, or another very old till are present in other parts of Jackson Hole and on the Gros Ventre Range. The higher remnants indicate that the glaciation was of such magnitude that ice containing erratics from the Absaroka Range moved 30 miles southwestward across the valleys of the Buffalo Fork and Gros Ventre Rivers, and on through several passes in the Gros Ventre Range at elevations above 10,000 feet. It is possible that the superimposed course of the ancestral Hoback River across the Hoback Range may have been controlled by the western margin of this ice mass riding across the red conglomerate before excavation of the Hoback Basin.

Depositional Event 19. HUCKLEBERRY RIDGE TUFF (UPPER PLIOCENE). The second and largest of three major explosive eruptions of rhyolite welded tuff in central Yellowstone National Park. This ash flow tuff engulfed the northern end of the Teton Range and flowed down both sides; at least 25 miles on the east side and 45 miles on the west. The K-Ar age has been recalibrated to 2.1 million years so it is now considered to be Pliocene rather than Pleistocene.

Tectonic Event 17. This is not just a single event but an on-going one. It began after emplacement of the Huckleberry Ridge Tuff. Total displacement near the north end of Jackson Lake has been more than 5,000 feet, with the Teton Range rising and Jackson Hole sinking. Farther south, lack of data on the mountain block precludes precise measurements. Extensive movement continued along the Hoback fault in southern Jackson Hole. The valley block dropped (or the mountain block uplifted, or both) at least 1,000 feet. Talus breccia of Paleozoic limestone fragments that accumulated along the rising fault scarp and that had a steep initial dip westward was rotated enough to make it dip eastward into the fault.

Glaciation Event 2. The second major glaciation was much smaller than the first and is much more clearly documented. It has been called by various names (Buffalo, Bull Lake, Pinedale). The ice had sources in Yellowstone National Park and in the Absaroka, Teton, Gros Ventre, and Wind River Ranges. The ice streams coalesced in Jackson Hole, where the thickness was as much as 3,000 feet. From there the ice flowed southward along the Snake River Valley, through the Snake River Canyon, and into Idaho.

Despositional Event 20. LEIDY FORMATION (PLEISTOCENE). Between 500 and 1,000 feet of fine-grained lacustrine and fluviatile strata intertonguing with granitic debris from the Wind River Range was deposited along a 30-mile-long lake that occupied the Gros Ventre River Valley. The outlet of the lake was plugged by southward-moving ice in Jackson Hole. This ice was apparently from one advance in Glaciation 2.

Glaciation Event 3. The third major glaciation, sometimes called Bull Lake, was only about half as extensive in Jackson Hole as the second, but the actual distribution is a subject of disagreement.

Glaciation Event 4. The fourth and last major glaciation is generally called Pinedale. Its extent is a matter of controversy. Some workers equate it with part of Glaciation 2, whereas others consider its down-valley progress is marked by the Burned Ridge moraine that crosses the middle of Jackson Hole. The moraines surrounding the lakes at the base of the Tetons all are part of this event.

Holocene Events. The Holocene has been a lively time in Jackson Hole. Less than 10,000 years ago the valley of Flat Creek at and north of the town of Jackson was downfaulted 150 feet. There was renewed movement on the Teton and Hoback faults. The outwash plain deposited south of the Burned Ridge moraine of Pinedale age, perhaps 15,000 years old, tilted west, possibly at a rate between one foot in 60 to one foot in 300 years. This faulting has caused major drainage changes in Flat Creek and the Snake River.

This faulting must have been accompanied by major earthquakes. These shook the glacially and tectonically oversteepened hillsides, many of which were composed of plastic strata, and created large and small landslides such as the Lower Gros Ventre slide that will be visited on this trip.

DAY 7

The trip backtracks the last part of Day 6 for 12 miles to the Moose junction. The geology along this route is complex and has been described in detail (Love, and Love, 1983) so only a condensation is presented here. On the west side of the highway (the east face of East Gros Ventre Butte) is a section of Ordovician, Devonian, Mississippian, Pennsylvanian, and Permian rocks, tilted westward, and unconformably overlain by gray basalt, scoriaceous red and black basalt, and brown andesite. These are some of the eight sequences of igneous rocks on the East and West Gros Ventre Buttes. Unfortunately, none has been dated (samples were submitted in 1969) but are probably middle to late Cenozoic.

The highway for the first four miles north of Jackson traverses a normal fault, at least part of which is less than 10,000 years old, with the valley block dropped 150 feet (see Holocene Events in summary). The result is a classic example of misfit drainage. Flat Creek, which flows southward through Jackson is a small stream in a broad open valley, whereas the Gros Ventre River, on the upthrown block a mile farther north is a major river on an upland plain.

Blacktail Butte, east of the Moose road junction, is a fault block of Paleozoic rocks on the east and north sides and white Miocene tuff and limestone on the west and south sides. A brief unnumbered stop is planned at Grand Teton National Park Headquarters so that the field trippers can buy pertinent geologic and natural history maps and publications. From Moose, the route goes north to the one-way road to String Lake and south to Jenny Lake.

Stop 25

This is a photo and geomorphology stop. The Quaternary features that have been passed and those that lie ahead are discussed in the road log referenced above and in the G.S.A. Centennial Field Guide (Love, 1987, p. 173-178). Jenny Lake is 225 feet deep and is surrounded by a classic example of the youngest moraines in the region (Glacial Event 4). This moraine emerged from Cascade Canyon which is a U-shaped valley directly west of the stop. Holocene fault scarps, with the valley block dropped as much as 180 feet in one instance, are visible both to the northwest and southwest of Jenny Lake.

To the south of Cascade Canyon is Mount Teewinot (elevation 12,325 feet), which is composed chiefly of dark gray layered gneiss and migmatite, at least 2,800 million years old. The Grand Teton is barely distinguishable from this site but a better view can be had farther south along the highway. This is the second highest peak in Wyoming (13,770 feet) but is by far the most spectacular (Figure 64). Teton glacier, on the northeast side of the peak, is the largest active glacier in the range. The peak is composed of the Mount Owen Quartz Monzonite (late Archean, about 2,500 million years old; Reed, and Zartman, 1973).

The route backtracks to Moose. If buses are permitted on the Moose-Wilson cutoff, shown by the dotted heavy line on Figure 60, the route will go directly to Teton Village, but if not, the route will go back to Jackson, west toward Wilson, and then north to Teton Village, as is indicated by the solid black line. From Teton Village, the tram will take the group to the top of Rendezvous Mountain.

Stop 26

The lower part of the tram ride affords a good view of the Archean layered gneiss (older than 2,800 million years), then in regular succession, the Middle and Upper Cambrian units: Flathead Sandstone, Wolsey Shale, then the lowest big cliff (Death Canyon Limestone), then the green Park Shale, and finally, the Gallatin Limestone. The next big cliff-former is the Ordovician Bighorn Dolomite. The top of the tram is on the Devonian Darby Formation. At the overlook east of the top tower, the flat ledges of Darby dolomite have fossil corals on the surfaces. The

FIGURE 64 Air oblique view west showing the Grand Teton and adjacent peaks, the Teton glacier, and its terminal moraine. To the left of the Grant Teton, in the background, are layered Paleozoic rocks dipping west on the west flank of the Teton Range. The checkered farmland in the distance is in the Teton Basin (Pierre's Hole) in Idaho. This basin is underlain by Miocene to Pleistocene fluviatile, lacustrine, and volcaniclastic rocks more than 2,000 feet thick (Love, and others, 1978). Photograph by Austin S. Post, University of Washington, Aug. 19, 1963.

view southwest from the tram is shown in Figure 65. This sequence is described and illustrated by Love, and others (1971, Figures 31 and 32, and Table 3), and, except for the Madison Limestone, is the section seen close-up from the tram.

From this stop the view north shows the asymmetry of the Teton Range and the normal faulting of the Paleozoic section in Granite Canyon in the foreground. The distant panorama of Jackson Hole to the northeast, east, and southeast gives an overview of (1) the behavior of the Snake River in response to local and regional tilting of the valley floor; east out of Jackson Lake, then southwestward to a point south of Wilson, as a result of westward tilting from movement on the Teton fault, then diversion southeastward in response to eastward tilting of the valley floor in southern Jackson Hole, caused by movement on the Hoback normal fault; (2) the fault block buttes of Paleozoic rocks, rising above the flat valley floor, remnants left by the uneven collapse of the ancestral Laramide Teton-Gros Ventre arch; (3) Cenozoic volcanic rocks capping the buttes, and involved in the faulting; (4) the ice-sculptured landscape done by a southward-moving ice stream 2,000-3,000 feet thick during Glacial Event 2; (5) the faulted-off west end of the Gros Ventre Range that once connected, unbroken, with the ancestral Teton Range during Laramide time.

FIGURE 65 View southwest from the top of the Teton Village tram showing the Paleozoic sedimentary section. National Park Service photograph by W.E. Dilley and R.A. Mebane.

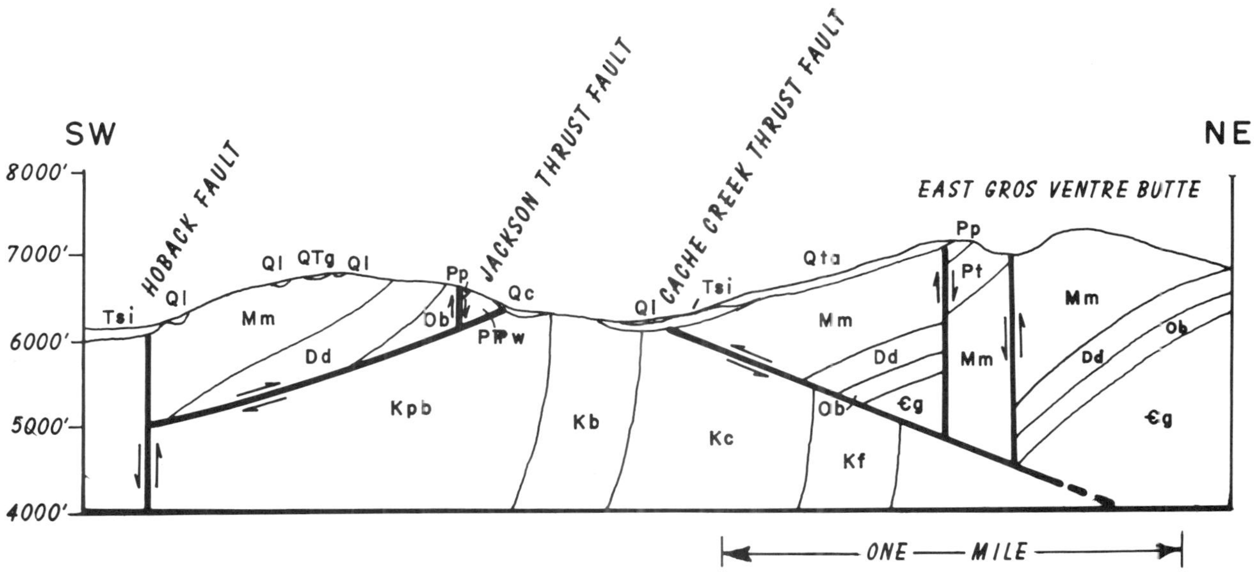

FIGURE 66 Northeast-trending cross section through the southwest part of East Gros Ventre Butte showing the relation of the Jackson and Cache Creek thrust faults a mile southwest of Jackson (after Love, and Albee, 1972). Symbols are: Qta – talus, Qc – colluvium, Ql – loess, all of Quaternary age; QTg – basalt scoria of Quaternary or Tertiary age; Tsi – Shooting Iron Formation (Pliocene); Kpb – post-Bacon Ridge strata (Cretaceous); Kb – Bacon Ridge Sandstone, Kc – Cody Shale, and Kf – Frontier Formation, all of Cretaceous age; Pp – Phosphoria Formation (Permian); IPt – Tensleep Sandstone (Pennsylvanian); IP Pw – Wells Formation (Permian and Pennsylvanian); Mm – Madison Limestone (Mississippian); Dd – Darby Formation (Devonian); Ob – Bighorn Dolomite (Ordovician); €g – Gallatin Limestone (Cambrian).

From this stop, descend on the tram and return to Jackson.

Jackson, Wyoming, Overnight Stop

DAY 8

The first thing in the morning the group will go south in Jackson to the overlook on the east side of the Americana Snow King Inn for orientation concerning the

Laramide and post-Laramide structures east and west of Jackson. Figure 66 is a cross section (Love, and Albee, 1972) of the buttes and valley one mile southwest of Jackson. The overlook is on vertical Cody Shale (below the mantle of Pleistocene loess), overridden from the south by the Jackson thrust fault and overridden from the north by the Cache Creek thrust. The view east up Cache Creek shows the same structural relationships. Data from the head of Cache Creek show that the Cache Creek thrust is slightly younger than the Jackson, and overrides it in several places. Both, however, are of Laramide age. The Jackson thrust was emplaced during Tectonic Event 7 and the Cache Creek during Tectonic Event 8. The faulted east face of East Gros Ventre Butte shows the steepening downward of Devonian, Mississippian, and Pennsylvanian rocks into the Cache Creek thrust. The dark scree streaming down from the top of the south end of the butte is black and red basalt scoria, the youngest of the volcanic rocks on the buttes.

After this orientation, the group will go northward through Jackson on U.S. Highway 26-87-189 for about eight miles to a junction just north of the Gros Ventre River, then northeast past the village of Kelly. This part of the route is described in detail elsewhere (Love, and Love, 1988). Kelly Warm Spring is on the south side of the road 1.5 miles north of Kelly. This spring flows an estimated five million gallons per day of clear, slightly sulfurous water with a maximum temparature of 86°F (30°C). Bubbles of gas, almost 90% nitrogen (Kvenvolden, and others, 1989), constantly agitate the pool surface. The water comes up along a Pleistocene normal fault, and was probably heated by the geothermal gradient at depth in the buried Madison Limestone aquifer. This is a favorite public place for bathing and kayak training. It also supports a variety of aquatic fauna and flora.

Stop 27

This stop is at the exhibit area on the Lower Gros Ventre Slide (Figure 67). It occurred on June 23, 1925, on the site of older slides. The debris was emplaced rapidly, roaring down the mountain and crossing the Gros Ventre River, perhaps in about three minutes according to eye witnesses. It apparently was triggered by an earthquake, but not as violent as that which caused the Hebgen landslide (at Stop 21). The momentum of the slide carried

FIGURE 67 Air oblique view west along Lower Slide Lake. Indicated are: Lower Gros Ventre Slide (A); Stop 27 (B); Devils Elbow Slide (C); type section of the Teewinot Formation (Miocene) (D); East Gros Ventre Butte (E); West Gros Ventre Butte (F); Teton Pass Stop 29 (G); Blacktail Butte (H); mud slide of Jurassic and Cretaceous shales (I); Cloverly Formation (Lower Cretaceous) (J); Nugget Sandstone (Jurassic or Triassic) (K); Red Slide that almost bisects lower Slide Lake (L). Photograph by J.D. Love, October, 1965.

debris as much as 350 feet up the north side of the valley and created a lake about 200 feet deep and three miles long. On May 18, 1927, the top 50 feet of the dam washed out and lowered the lake that much, causing a flood downstream. A wall of water 15 feet high raced through the village of Kelly, which at that time had 80 inhabitants. Six people were drowned. The Snake River, below its junction with the Gros Ventre River locally rose as much as 50 feet and the town of Wilson, which will be seen later in the day, was flooded by six feet of water. Voight (1978) considered this to be the largest (50 million cubic yards) slide of historic record in the United States.

The pink shales along the sides of the slide are in the Amsden Formation (Pennsylvanian and Mississippian), which is capped by the hard gray Tensleep Sandstone (Pennsylvanian). The rocks dip northward at about the same angle as the surface slope. The river undercut the slab of Tensleep which was perched there, unsupported, overlying the plastic shales in the Amsden Formation that were lubricated by melt water from the heavy snow pack of the preceding winter. Conditions therefore, were just right for a slide when the earthquake occurred.

The route goes eastward along Lower Slide Lake and numerous slides can be seen on both sides of the river. About a mile southeast of the upper end of the lake, the road swings around a point on a high narrow dugway and straight ahead is a magnificent view of the Red Hills anticline (Figure 68). Depending on road conditions, this may or may not be a safe place to stop. All the rocks are part of the Chugwater Formation which here is about 1,200 feet thick. The light-colored layers within the red siltstones and sandstones are gypsiferous siltstone. They emphasize the rollover of strata from the gentle northeast flank to the vertical southwest flank. There may be a thrust fault concealed under alluvium along the southwest flank. Two dry holes were drilled to Mississippian rocks. This anticline is a Laramide fold of pre-Tertiary age, considerably older than the Gros Ventre Range which lies at an oblique angle to the south. The age of folding is known because the Pinyon Conglomerate, which is all Paleocene in this part of Jackson Hole extends unfolded across the north end of the anticline.

The dugway here is cut into the Leidy Formation of Pleistocene age. It is a glacio-fluviatile-lacustrine sequence about 440 feet thick. The upper half is chiefly reddish-brown to gray very fine-grained clay, whereas the lower part is more sandy and conglomeratic. A till in the middle contains boulders of granite from the Wind River Range, at least 45 miles to the southeast. The route continues past the Red Hills anticline and more magnificent exposures of the red Chugwater Formation along the Gros Ventre River. The end of this trip is about one-fourth mile east of Crystal Creek.

Stop 28

From this stop the views are as follows: To the northwest (Figure 69) is the Lavender Slide which is composed chiefly of Jurassic and Lower Cretaceous shales. Parts of it are still moving slowly toward the river, crowding it southward in an incised channel at the near edge of the slide. Figure 69 identifies some of the key bedrock units. This slide is of some economic significance, for at one

FIGURE 68 View east at Red Hills anticline, a pre-Paleocene Laramide fold. At its northwest end, unfolded Pinyon Conglomerate, which here is entirely Paleocene in age overlaps the anticline. All rocks in this view are part of the Triassic red Chugwater Formation. Photograph by J.D. Love, August 26, 1955.

FIGURE 69 View northwest at the "Lavender Slide" on the north side of the Gros Ventre River. The slide debris is derived chiefly from Jurassic and Lower Cretaceous rocks. Parts of it are still moving slowly toward and into the river, crowding it southward in an incised channel at the near edge of the slide. Bedrock units indicated are: Triassic red Chugwater Formation (A); Jurassic marine gray Sundance Formation (B); Lower Cretaceous variegated member (hence the name "Lavender Slide") of the Cloverly Formation (C); Lower Cretaceous "Rusty Beds" member of Cloverly Formation (D); sandstones in the Upper Cretaceous Frontier Formation (E). Photograph by J.D. Love, August 26, 1955.

brief time it dammed the Gros Ventre River. Behind the dam, gold-bearing quartzite clast gravel accumulated. The gravel came from more than 1,000 feet of Pinyon Conglomerate that caps the three Leidy peaks (from west to east, Mount Leidy, elevation 10,326 feet and the lesser peaks, Middle Leidy, and East Leidy). A major electrical company, in 1965, invested more than a million dollars in a gold placer recovery machine that was not capable of recovering the "flour gold" (dandruff-size flakes). The enterprise was eventually abandoned. The same company installed a similar gold machine with identical results in the Snake River Canyon where even thicker placer deposits are present. The route will pass this site on Day 9.

Northeast and east of Stop 28, the high gray hills across the Gros Ventre River are spectacular outcrops of sandstone and shale, the Mowry Shale (Lower (?) Cretaceous) and the overlying Frontier Formation (Upper Cretaceous). To the south is a view up the glaciated valley of Crystal Creek. Lateral moraines of Glaciation 4 are prominent on both sides of the valley. Where Crystal Creek emerges from its steep-walled canyon on the north flank of the Gros Ventre Range, excellent exposures of Mississippian to Jurassic strata are present, especially on the east side (Simons, Love, and others, 1988).

West of the stop, the red butte rising nearly 800 feet above the valley floor, is composed of the red Chugwater Formation. Plastered against the lower 650 feet of mountainous topography on the east face of the butte are plastic clays of the Leidy Formation (Depositional Event 20). The middle part of this exposure has slid down in a conspicuous dumbbell-shaped landslide which is still moving. In the last few years it has reached the main road at the Crystal Creek bridge and is causing maintenance problems.

Retrace the route back to Jackson and turn west on U.S. Highway 189, then 1.5 miles west of town, turn right at the "Y" on State Highway 22. The geology of this area has been mapped in detail (Love, and Albee, 1972). In a quarry face on the north side of the highway one-half mile west of town, is a gray basalt of unknown age but probably mid- to late Cenozoic (samples were submitted for age determination in 1969). The basalt was intruded along the Cache Creek thrust fault (Tectonic Event 15) where it overrides vertical Upper Cretaceous strata.

At the "Y", the route is on the sandwich of vertical Cretaceous strata (mantled by Quaternary loess), with Paleozoic rocks of the Jackson thrust plate overriding them northward (Tectonic Event 7) and Paleozoic rocks of the Cache Creek thrust plate overriding them southward (Figure 66, Tectonic Event 8). On the west side of the Paleozoic strata on West Gros Ventre Butte, is an undrained graben of Holocene age in loess and swamp deposits. This is another example of the recency of tectonic movement on the floor of Jackson Hole.

At the Snake River bridge are outcrops of greenish gray porphyry, full of inclusions of mafic Precambrian rocks. Again, its age is not known, but it is probably mid- or late Cenozoic. It was apparently intruded along a concealed normal fault system, and is close to the Cambrian-Precambrian contact. The adjacent Snake River at the bridge is approximately nine feet higher than Fish Creek at its highway bridge 1.5 miles farther west. Thus, the major south-flowing stream is not the lowest one in the valley. This is because of Holocene westward tilting of the valley floor here and farther north (Love, and Montagne, 1956). As mentioned in connection with the Gros Ventre Slide and subsequent flood, the discrepancy in elevation of the valley floor is why the village of Wilson was inundated with six feet of water during this flood. That was the last of several times that the village had been flooded by the Snake River. Now that flood control levees have been installed along the river, the question arises as to whether the valley floor will tilt westward faster than the levees can be built up.

On the west side of Wilson is a fault scarp more than 100 feet high that displaces Quaternary deposits (Love, and others, Grand Teton National Park geologic map, in press). The course of Fish Creek, parallel to, and close to the fault, probably was controlled by this downdropping.

The route passes around the south end of the Phillips Canyon fault block composed of west-dipping Paleozoic rocks. This is the fourth of these west-dipping faulted remnants of the same Paleozoic rocks on the collapsed Laramide Teton-Gros Ventre arch across this part of Jackson Hole (from east to west, Miller Butte, East and West Gros Ventre Buttes, and Phillips Canyon block). Where the highway turns north for one-half mile, there is a huge road cut on the right, exposing gray andesite which has a K-Ar age of eight million years (Tectonic Event 15). At its base, in contact with the Madison Limestone, is a white crystal tuff bed that is baked red near the contact with the overlying andesite. Thus, it appears that there was first an eruptive phase of volcanism, then the extrusion of the andesite along the Phillips Canyon normal fault. This was a major fault, along which the Phillips Canyon block was dropped about 4,500 feet below the main part of the Teton Range (Schroeder, 1972).

Around the next bend to the west is the best fresh exposure of the Middle Cambrian Flathead Sandstone in the region. Up the notorious Glory Mountain snow avalanche chute that bisects this outcrop, is a regular section of Paleozoic rocks up to the Amsden Formation on top of this part of the Teton Range.

Stop 29

Teton Pass (elevation 8,431 feet) is at the right angle juncture of the northwest-trending leading edge of the Thrust Belt (the mountain ranges of which have no Precambrian cores) and the south-southwest-trending Teton Range (a Foreland range with a Precambrian core; see Figure 60). At this stop is the red Chugwater Formation. The Tensleep Sandstone is on the north and Cambrian limestone on the south. Because of its complexity and accessibility, this has been the site of many studies. Figure 70 is a detailed interpretation favored by Dunn (1983) who also presented several others. Additional ones were made by Zeller (1972), Schroeder (1969) and Oriel and others (1985). All but the last one showed both the

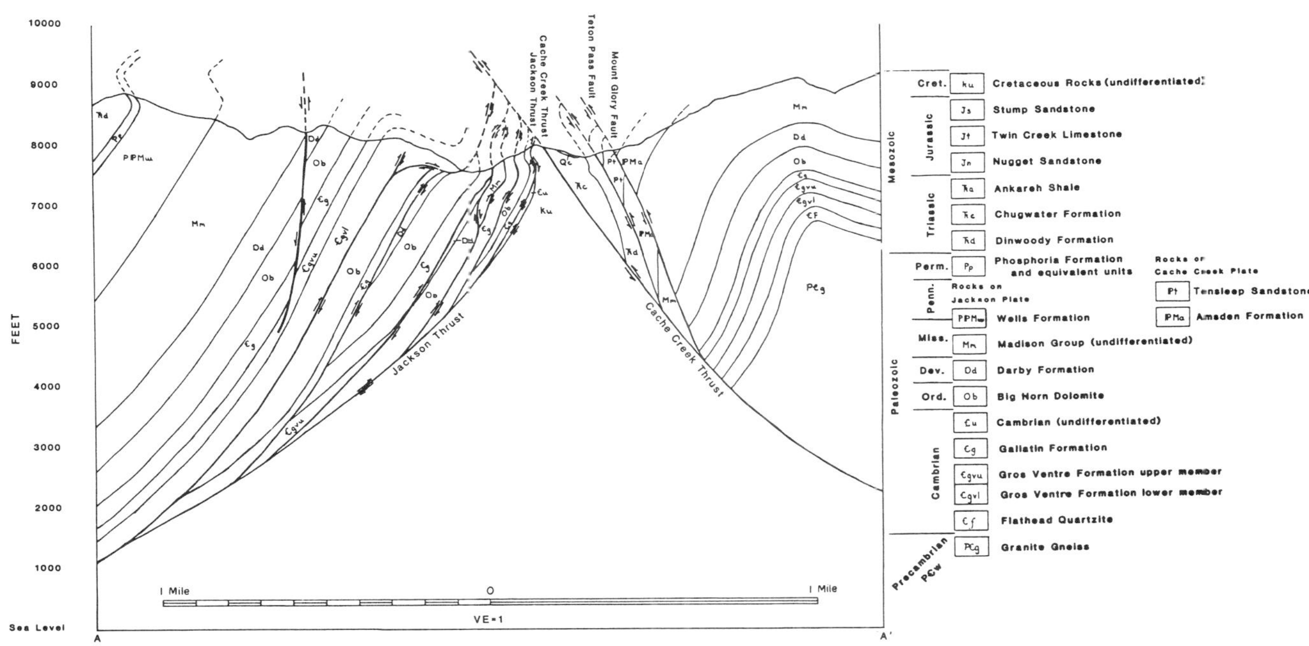

FIGURE 70 Cross section from south (left) to north through Teton Pass showing the interpretation of thrust structures favored by Dunn (1983).

Jackson and Cache Creek thrusts extending through the pass and continuing northwestward into Idaho. The structure of the pass has been complicated by multiple thrust slivers of Paleozoic rocks related to the Jackson thrust on the south side and by several faults on the north side that have been interpreted by some workers as normal faults and by others as thrusts, depending on the concept of timing of emplacement of each sliver. The amount of displacement on the Cache Creek thrust here is probably only a very few miles, whereas that on the Jackson thrust may have been 50 miles or more. The sedimentary facies of both Paleozoic and Mesozoic rocks on the opposing thrust sheets is very different. There is no disagreement as to the presence of Upper Cretaceous strata in the sandwich between the faults to the northwest down Trail Creek, nor is there to the east in Jackson Hole.

Fresh exposures of the vanadiferous and phosphatic black shales in the Phosphoria Formation are visible in several road cuts on the west side of the pass. Another interesting occurrence is copper, silver, and zinc in the Nugget Sandstone, which seems to have an affinity for these elements in many places for the next 100 miles to the south (Love, and Antweiler, 1973).

The view east from Teton Pass shows many geologic phenomena not visible from Stop 26 at the top of the Teton Village tram. For example, the scene southeastward in a nearly straight line for 15 miles to and beyond Jackson and up the valley of Cache Creek shows the continuity of the opposing thrust sheets, and in between, the sandwich of nearly vertical Cretaceous strata overridden from both sides. Out of sight from this view, the thrusts diverge, with the Cache Creek continuing southeastward along the southwest flank of the Gros Ventre Range and the Jackson turning southward and possibly becoming the Prospect thrust of areas to the south. This is also a vantage point to get a concept of the magnitude of the ice mass (or masses) that filled Jackson Hole. As can be seen along the route to Teton Pass, southeastward-moving ice from the Tetons filled this part of Jackson Hole to about 400 feet below the pass. At Wilson the elevation is 6,140 feet so the ice was about 2,000 feet thick at Wilson. Ice went over the Gros Ventre Buttes and over Snow King Mountain, the one with the ski runs on it directly south of Jackson. A conspicuous south-trending U-shaped hanging valley can be seen two miles southwest of Jackson and 1,000 feet above the valley floor.

Return to the "Y" and go south on U.S. Highway 26 for 12 miles to Hoback Junction. The route parallels the Hoback normal fault (Tectonic Event 13), with movement opposite that of the Teton fault (the Hoback is hinged on the west and the block tilted eastward). At the first Snake River bridge south of Jackson, the view east shows the man-made Squaw Creek landslide that buried the old highway. This slide was activated by highway construction many years ago and after expenditure of several hundred thousand dollars and frequent slide blockages, the highway was rerouted to its present position on the west side of the river.

At Hoback Junction, the route turns east on U.S. Highway 187-189. The Willow Creek anticline in gently dipping Lower Cretaceous black shale and gray sandstone is along the south side of the Hoback River. The gray cliffs to the east are the basal conglomerates of the Miocene Camp Davis Formation (Depositional Event 14). It is more than 5,500 feet thick and was derived from the Gros Ventre Range to the northeast. After deposition, it was down-faulted along the Hoback normal fault. Geophysical data indicate that this is a listric fault (Figure 71) and cross section reconstructions north of the Hoback Canyon suggest that in places it may have as much

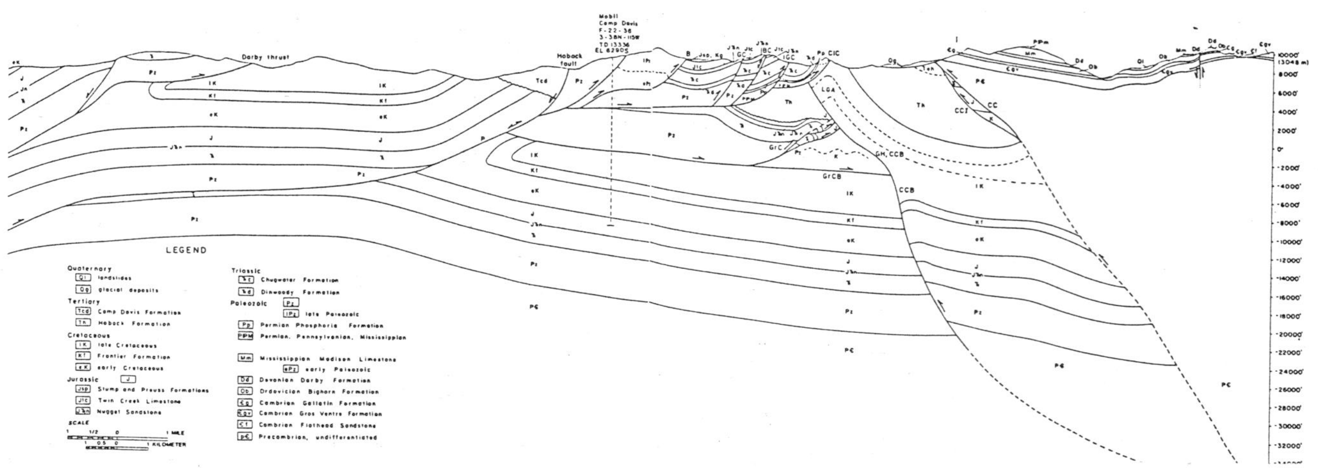

FIGURE 71 Cross section through the Stinking Spring, the Mobil well, past Little Granite anticline, and on to the southwest margin of the Gros Ventre Range. Abbreviations of major faults and anticlines, from west to east: P – Prospect (Jackson) thrust, B – Bear thrust, GC – Game Creek thrust, BC – Bull Creek thrust, IGC – Little Granite thrust, CIC – Cliff Creek thrust, GrC – Granite Creek thrust, LGA – Little Granite anticline, GrCB – Granite Creek blind thrust, CCB – Cache Creek blind thrust, GH – Game Hill thrust, CCI – Cache Creek subfault, CC – Cache Creek thrust. After R.B. Hunter, 1987.

as 10,000 feet of displacement. Masses of Mesozoic rocks are incorporated in, and overlie the upper conglomerates. Whether they are thrust masses or landslide blocks is a matter of dispute. If they are slide blocks, there is a question as to the source area. Not only is an area of adequate size hard to find, it will take a major amount of tectonic juggling to slide coherent masses of plastic strata downhill to the present area of the Camp Davis Formation.

A series of highly sulfurous springs emerges along the Hoback fault where it crosses the Hoback River. Pioneers called these the "Who-poop" Springs but the name has been laundered to "Stinking Springs". Despite the "Point of Interest" sign erected by the Wyoming Highway Department, saying that these are hot springs, the average temperature is 48°F (9°C). The combined flow of the springs above the low water stage of the river is estimated at about three million gallons per day of milk-colored water, cluttered with streamers of white sulfur. Other unmeasured flows can be seen emerging from the river bed (field notes by J.D. Love, Sept. 4, 1977).

About three-eighths of a mile east of the springs a deep dry test well was drilled by Mobil Oil Co. It started in the Madison Limestone and bottomed in Jurassic rocks at a depth of 13,336 feet. This well is useful in interpreting the Thrust Belt structure because it encountered three thrust faults that reach the surface farther east. The lowest and biggest is at a depth of 7,823 feet. Thus, the dips on the thrust planes and the strata involved can be determined in the subsurface (Figure 71). This section by Hunter (1987) goes northeast through the well and past the Little Granite anticline and on to the Gros Ventre Range. Figure 72 shows what the structure looks like on the surface.

The route continues up the Hoback Canyon for six miles and at its east end, turns sharply northeastward up Granite Creek. About half way along this part of the canyon is an asymmetric north-trending anticline in Paleozoic rocks on the overriding block of the Bear thrust fault which puts them on red Triassic rocks (Figure 73). This cross section shows the structure in the easternmost three miles of the canyon. A mile east of the Bear thrust is the Game Creek thrust that involves Jurassic rocks. Two miles still farther east, at the mouth of Granite Creek is the Cliff Creek thrust, the eastern edge of the Thrust Belt. Battle Mountain, on the right, is composed of Nugget Sandstone, thrust eastward onto vertebrate fossil-bearing gray sandstones and claystones of the Hoback Formation (Figure 74).

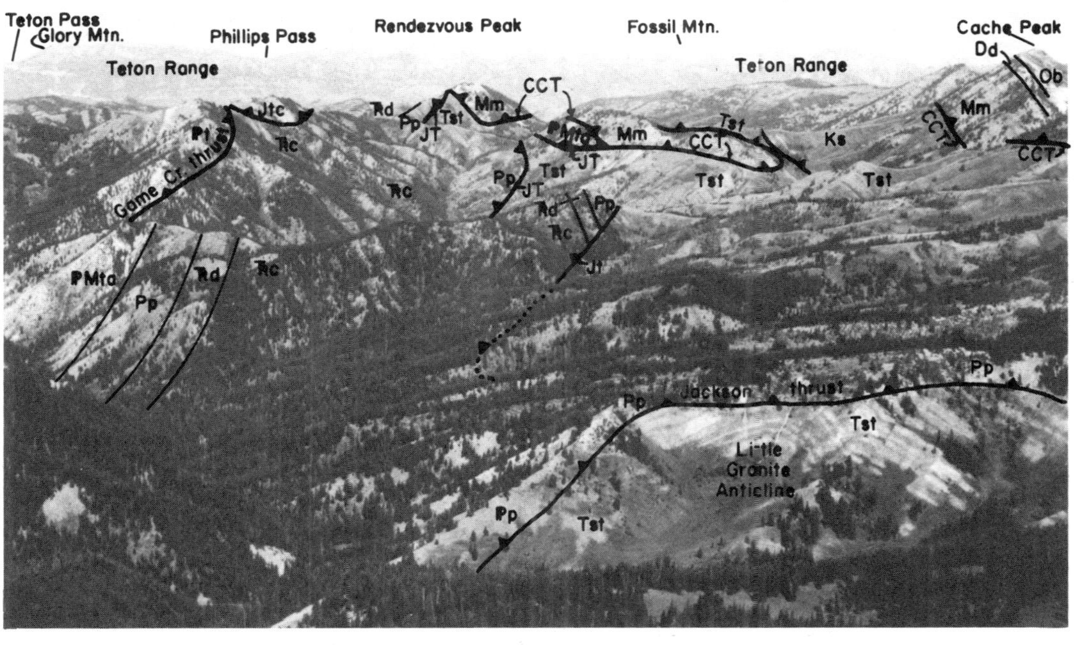

FIGURE 72 Air oblique view west from near head of Horse Creek, showing leading edge faults of the Thrust Belt on left, moved toward right, and Cache Creek (CCT) foreland thrust at upper right, moved toward left. Little Granite anticline in Hoback Formation is at lower right, overridden by Paleozoic rocks of the Jackson thrust plate (JT), and also overridden by Paleozoic rocks of the Cache Creek thrust plate at upper right. Formations indicated are: Ob – Bighorn Dolomite (Ordovician), Dd – Darby Formation (Devonian), Mm – Madison Limestone, IPMta – Tensleep Sandstone and Amsden Formation (Pennsylvanian and Mississippian), IPt – Tensleep Sandstone (Pennsylvanian), Pp – Phosphoria Formation (Permian), T̶d – Dinwoody Formation (Triassic), T̶c – Chugwater Formation (Triassic), Jtc – Twin Creek Formation (Jurassic), Ks – Sohare Formation (Upper Cretaceous), Tst – Skyline Trail Conglomerate Member of Hoback Formation (Paleocene and earliest Eocene). Photograph by J.D. Love and W.B. Hall, Aug. 25, 1971.

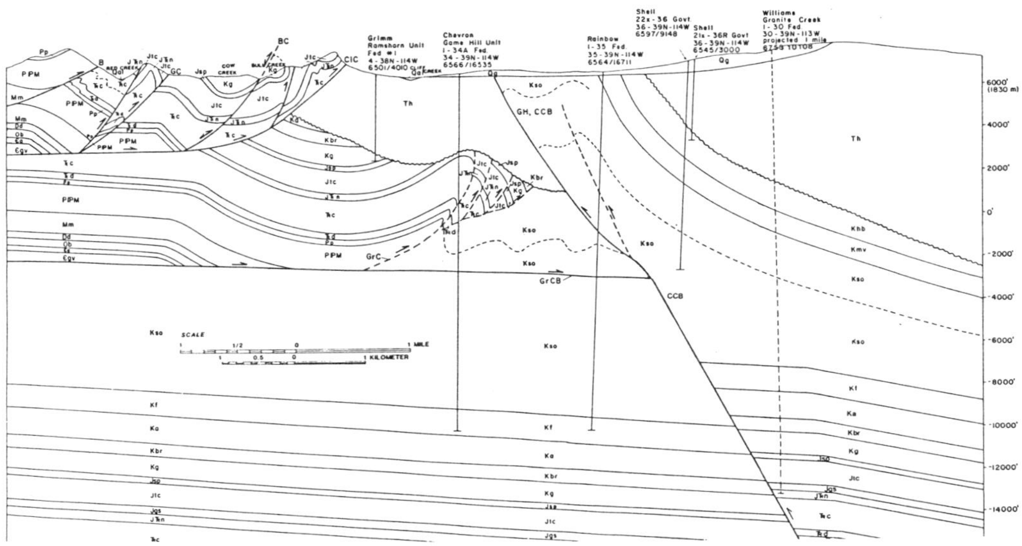

FIGURE 73 West to east cross section near the junction of Granite and Little Granite Creeks showing a structural interpretation from drilling and seismic data in the northwest part of the Hoback Basin that is traversed by the field trip route on the way to Stop 30. For formation symbols and abbreviations of major faults see Figure 71. After R.B. Hunter, 1987.

FIGURE 74 View north at Battle Mountain composed of Nugget Sandstone on the leading (eastern) edge of the Cliff Creek thrust, the easternmost one in the Thrust Belt. The Nugget overrides vertebrate fossil-bearing Paleocene part of the Hoback Formation which dips about 60° to the east. This view is about a mile south of the west-east cross section shown in Figure 73. Photograph by J H. Rathbone.

How and when did the Hoback River cut a canyon 1,000 feet or more deep across the Hoback Range from the Hoback Basin to the south end of Jackson Hole? Dorr, and others (1977) postulated headward erosion of ancestral Hoback River. The problem with that is that many resistant ridges of strata lie at right angles to any west-flowing stream and headward erosion through them would be difficult. Such a pattern encourages trellis drainage rather than the present almost straight course that is not diverted by the ridges of hard rocks. There is no evidence of antecedency. A plausible explanation is superposition, with the ancestral Hoback River established on the surface of the red conglomerate (Depositional Event 16) at an elevation of 7,500 to 8,000 feet, perhaps in late Miocene or Pliocene time before the Hoback Basin had been excavated. The Hoback Rim on the east side of the basin is at an elevation of about 8,000 feet and it seems probable that the basin was filled to about that level before regional excavation began. Also, ice from Glaciation 1 poured southward through passes in the Gros Ventre Range at elevations of 10,000 feet and above. Meltwater from this ice could have augmented the flow of the ancestral Hoback River on its high-level westward superimposed course established on the red conglomerate, and facilitated the cutting of the canyon.

The route up Granite Creek for the nine miles to Stop 30 is mostly on the Hoback Formation of Paleocene and earliest Eocene age. A concealed back thrust (one that moved from east to west), the Game Hill, also called "Battle Mountain fault" by Schroeder (1976), not to be confused with the Game Creek thrust farther west (Figure 73), results in repetition of part of the Hoback Formation. This formation here is about 6,000 to 10,000 feet thick and consists of nonmarine gray sandstones and shales. In the upper part is the red Skyline Trail Conglomerate, as much as 3,000 feet thick (Figure 72). Nearly all the clasts are derived from adjacent Paleozoic and Mesozoic rocks and are moderately rounded. Some boulders are as much as five feet in diameter. On the Little Granite anticline (Figures 71, 72) the base of the conglomerate is only a few hundred feet above Cretaceous rocks, whereas along Granite Creek to the southeast, it is at least 4,000 feet above the Cretaceous strata.

Stop 30

Granite Hot Spring emerges along the Cache Creek thrust fault. It has an average temperature of 106°F (41°C) and a flow of about 300 gallons per minute. When the spring is diluted by surface water and the flow increases, the temperature decreases. The aquifer is in Cambrian strata not far above the Precambrian rocks. Granite Falls, downstream from the swimming pool at the spring, is on the Flathead Sandstone. Rising for 2,200 feet above the spring is a series of awesome cliffs in a much-studied section of Cambrian, Ordovician, Devonian, and Mississippian rocks similar to that seen at Stop 26 near the Teton Village tram. To the east is the "Flying Buttress" composed of Madison Limestone curving down southward in steep ledges, dragged into the Cache Creek thrust. Beneath the thrust are poor exposures of the Hoback Formation. The amount of structural displacement is probably several miles. The depth of the Madison Limestone from its surface exposure here to its position under this part of the Hoback Basin is probably about 25,000 feet (Figure 73). To the west, the Cache Creek thrust is well exposed, with the Paleozoic rocks bending down and overriding the upturned edges of the Skyline Trail Conglomerate.

Return to Jackson. The return trip serves to give a slightly different view of the geology along the route and allows the opportunity for questions that may have arisen.

Jackson, Wyoming, Overnight Stop

Days 9 and 10 are devoted entirely to the Wyoming-Idaho-Utah Thrust Belt. This belt is frequently called the "Overthrust Belt". However, many of the thrusts are backthrusts or underthrusts so their significance is disguised if they are all lumped as "Overthrusts".

Much of the following presentation, especially beginning with Stop 33, was prepared by James C. Coogan, who has spent several years studying the Thrust Belt for his doctoral dissertation. In addition, he worked more years for the petroleum industry in that area. In this capacity, he had access to surface, subsurface, and geophysical data not generally available to other researchers.

These last two days are designed to demonstrate the interrelationships between the structure and stratigraphy of the Thrust Belt. Two structural episodes are well exposed in the north-trending mountains and valleys along the field trip route. The Cretaceous through Early Eocene thrust structures are displayed by ridge-forming Paleozoic strata that define the leading edges of the principal thrusts in the belt (Figure 75). Intervening valleys are structural half-grabens associated with Miocene to Holocene Basin and Range normal faulting. The stratigraphy of the region records three general depositional regimes: the Proterozoic through Jurassic miogeocline, the Cretaceous through early Eocene foreland basin, and the Miocene to Holocene intermontane rift basins.

The Snake River transect from Jackson to Alpine provides a view of several structual and stratigraphic associations that will serve as recurring themes for the remainder of the trip: (1) thrusts shorten a westward-thickening Paleozoic through Jurassic shelf and miogeocline sequence (Figure 76); (2) thrusts in the eastern half of the belt carry no strata older than the Cambrian Gros Ventre Formation and are detached along a sole decollement above crystalline basement; (3) a variety of cross-cutting, overlapping, and provenance stratigraphic criteria demonstrate that the main episodes of emplacement for the major thrust sheets are younger toward the foreland (Figure 77; Royse, and others, 1975; Wiltschko, and Dorr, 1983); (4) younger imbricate thrusts move farther toward the foreland because there is less frictional drag, especially where there is buttressing by the foreland that stops the older thrusts; and (5) Miocene to Holocene extensional faults localize at the position of earlier thrust ramps.

DAY 9

The first 13 miles of the route is a repeat of the last part of Day 8. From Hoback Junction, turn southwest on U.S. Highway 26-89 for four miles, going down the section of east-dipping Lower Cretaceous rocks. A detailed

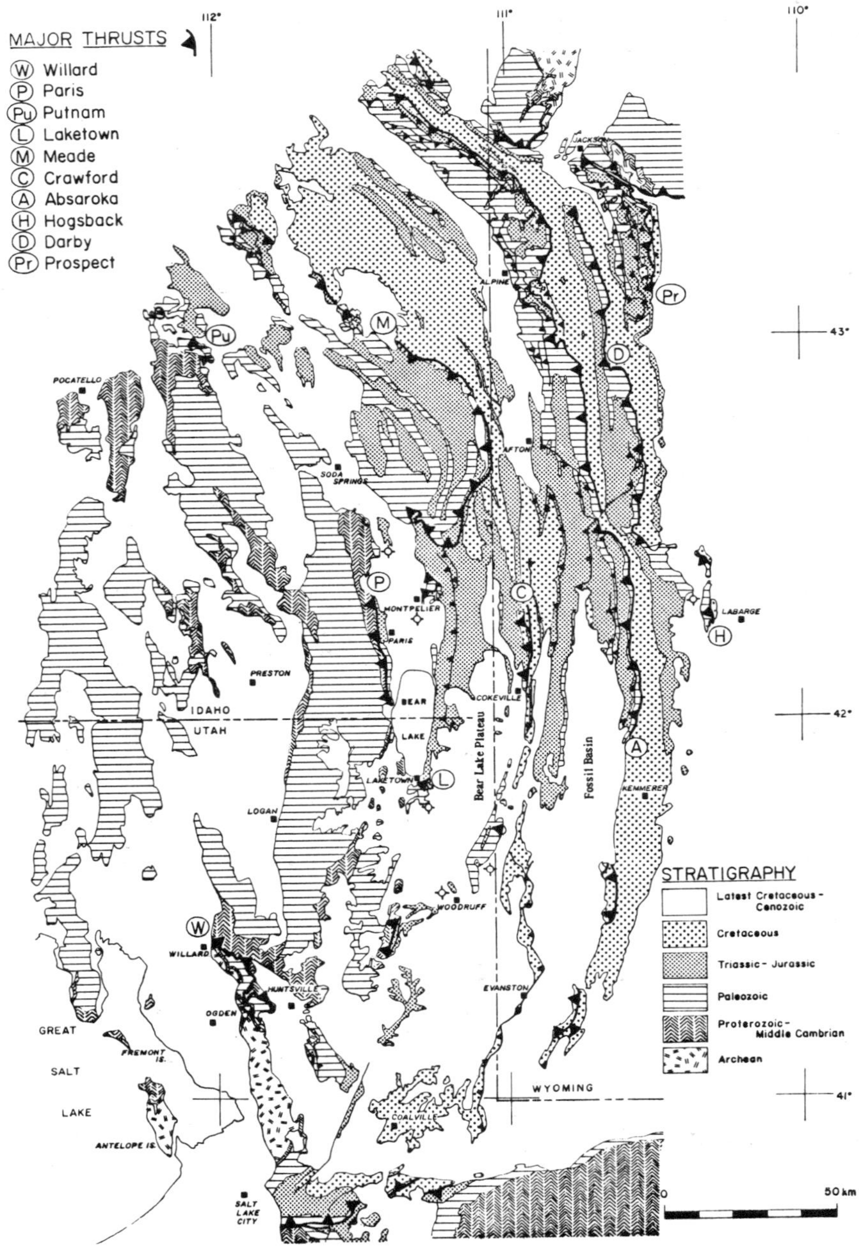

FIGURE 75 Generalized tectonic map of the Wyoming-Idaho-Utah Thrust Belt. Compiled by J.C. Coogan.

road log of the geology from Jackson to Alpine Junction is available elsewhere (Love, 1979).

Stop 31

Astoria Mineral Hot Springs, (at Stop 31, Figure 60) (formerly called Johnny Counts Warm Spring (Schultz, 1918), later, Ben Goe's Warm Spring, and finally in the 1960s, Astoria) emerge from a travertine-covered flat which straddles the crest of an anticline in the Mission Canyon Limestone (upper part of the Madison Group here). A ^{14}C age on shell material from an excavation on this flat is 11,940 ± 500 years (Meyer Rubin, written commun., 1959, Lab No. W-1070), so the 20 feet or more of travertine down to the present river level must be much older. The temperature in the main pool is about 104°F (40°C) but in the hot baths it is about 114°F (45°C).

Figure 78 shows the structure on the north side of the Snake River Canyon directly northeast of the hot springs. The main trace of the Darby thrust, which is one of the major thrust faults in the region, trends north up the small canyon and brings the sandstone of the Wells Formation (Pennsylvanian) over the Jurassic Twin Creek Limestone. The gray-green, non-resistant Twin Creek beds are then thrust over redbeds of the Jurassic Preuss Formation.

On the hills west of Astoria is a complexly folded section of Twin Creek to Wells on the hanging wall of the Darby thrust plate. The exposures at the east end of the highway cut on the north side of the Snake River are sandstones in the upper Wells. They are in normal fault contact with a complete section of the Phosphoria Formation. At the top is the thin black phosphatic shale of the Retort Member; the middle is the cliff-forming Rex Chert Member; and at the base is the black shale and phosphorite of the Meade Peak Member. This is one of the classic sections of the Phosphoria Formation (measured by Eliot Blackwelder in 1910, *in* Schultz, 1918). Both of the black shale members contain a variety of trace elements, one of which is uranium. In the 1950s, Ben Goe, who owned the hot springs at that time, dug an adit into the Meade Peak black shale at the highway cut and installed benches on which people who were not feeling well could come and get irradiated at 50 cents per hour.

Around the bend from Astoria, on the east side of the Snake River, is a gravel deposit 200 feet thick, known in gold placer mining circles as Pine Bar. A second and thicker (240 feet) gravel deposit four miles farther downstream is known as Baileys Bar. Both are remnants of a vast accumulation of gravel behind a landslide dam. The slide debris flowed northwest out of the valley of Bailey Creek and dammed the Snake River, backing up a lake nearly as far as Jackson. The gold originated in the same gold-bearing quartzite clast conglomerates of Cretaceous and Paleocene age as those north of Stop 28. The gold was carried down by tributaries to the Snake River which then took it to the dam site. When the dam was emplaced, the gold and gravel settled out, just as it did along the Gros Ventre River at Stop 28. Gold at Pine Bar and Baileys Bar has been known since 1870 and the deposits and pioneer attempts at exploitation were described in detail by Schultz (1914). The gold is "flour gold", the flakes micro-dandruff in size, and are not readily recovered by conventional nugget-recovery methods. Futile attempts at recovery have been made during the last 100 years. The final one was by the same company that tried it at Stop 28 on the Gros Ventre River. The U.S. Forest Service ordered the bulky recovery equipment removed so the operator tried to ferry it across the river. The boat capsized and the equipment is at the bottom of the river.

Stop 32

Four miles downstream from the right-angle bend in the river at the mouth of Bailey Creek, the route crosses the Greys River anticline. Its eastern flank consists of overturned Jurassic and Lower Cretaceous rocks; the west flank is more gentle and has very good exposures, especially of the black and gray shales and sandstones in the Bear River and Aspen Formations of early Cretaceous

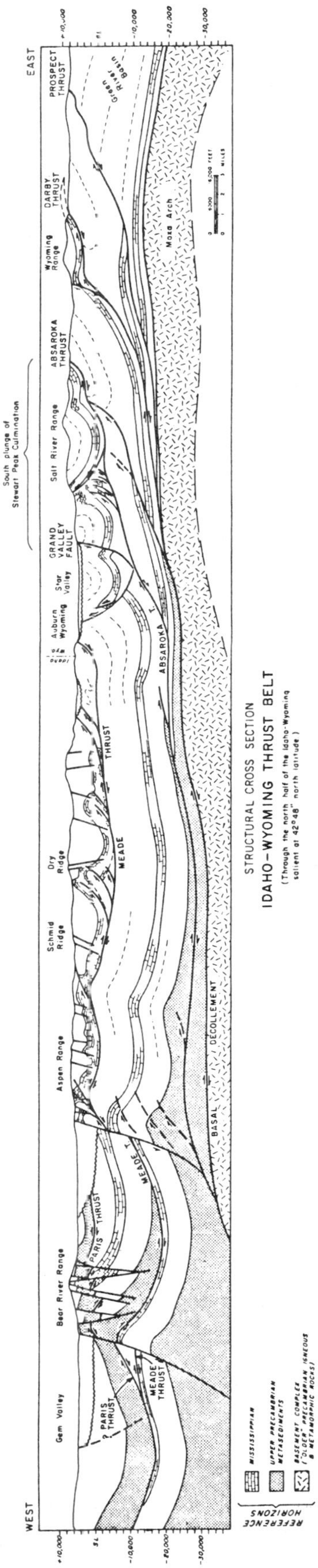

FIGURE 76 Generalized cross section of the Wyoming-Idaho part of the Thrust Belt at Lat. 42° 40′. From Lageson (1980 after Royse, and others, 1975).

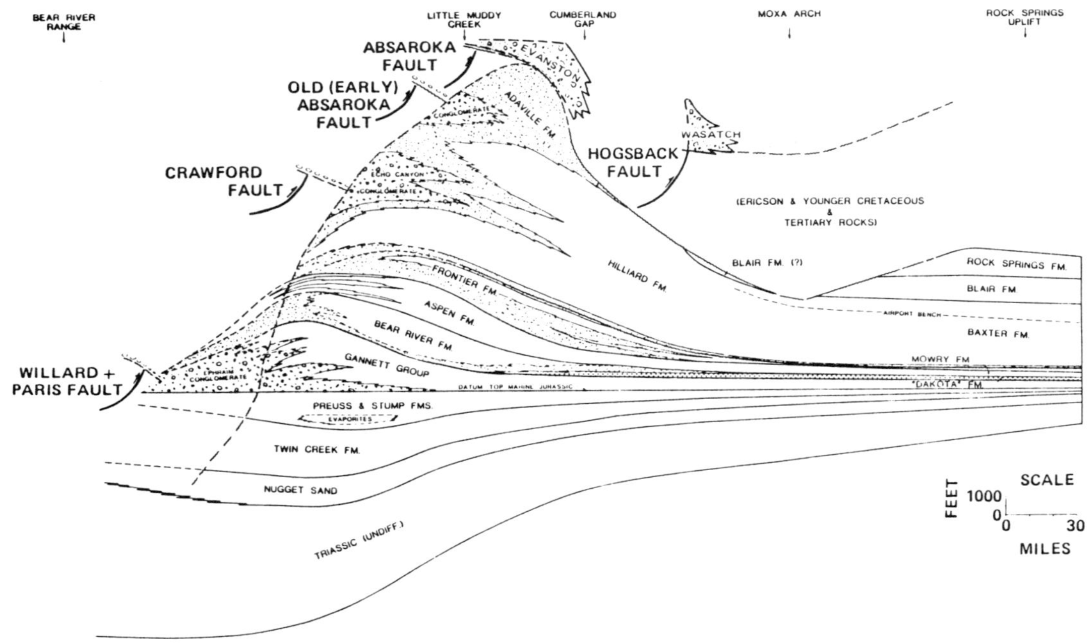

FIGURE 77 Diagram illustrating the relationship between Foreland Basin sedimentation and thrust source terrains in the Wyoming-Utah part of the Thrust Belt. Modified from Lamerson (1982) after Royse, and others, (1975).

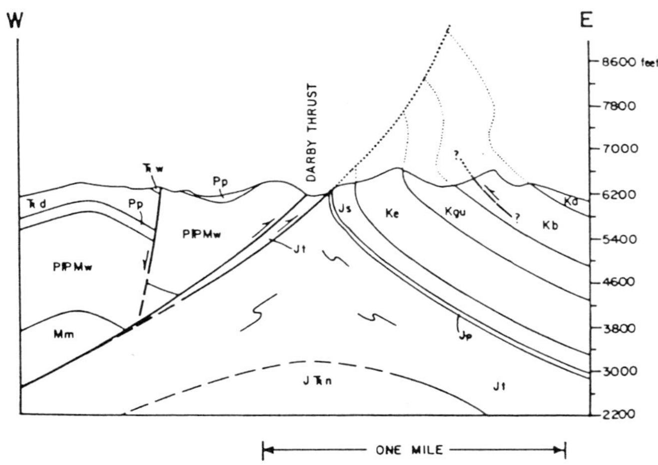

FIGURE 78 A west-trending cross section located 1,100 feet north of the Snake River across from the Astoria Mineral Hot Springs, showing the Darby thrust fault and adjacent structure. Projections of bedding attitudes along strike from the north and south were used in the reconstruction of contacts east of the Darby thrust. Geology after Albee (1968).

age. Overriding the Greys River anticline from the west is the gray wall of Madison Limestone on the hanging wall of the Absaroka thrust plate. This thrust is one of the largest and longest in the Thrust Belt.

Stop 33

The canyons of the Snake and Greys Rivers east and southeast of Alpine expose the deepest structural and stratigraphic levels in the Absaroka thrust system. At Alpine, imbricates of the Absaroka thrust include the St. John and Ferry thrusts. The St. John thrust is well exposed on the north wall of Snake River Canyon (Figure 80) where the thrust places Ordovician Bighorn Dolomite over a folded and imbricated Mississippian through Pennsylvanian section. A cross section along the north side of Greys River Canyon (Figure 81) demonstrates that the Ferry, St. John, and Absaroka thrusts all share a decollement in the Cambrian Gros Ventre Formation. Thrust dips steepen westward as the early imbricate, Ferry thrust, is back-folded by later movement on the St. John and Absaroka thrusts, respectively. This folding of thrusts illustrates the progression of imbrication toward the foreland that is the general rule throughout the Wyoming-Idaho-Utah Thrust Belt.

An exception to this rule is displayed by the Firetrail thrust in cross section and down-plunge map view (Figure 81). In cross section, the Absaroka thrust is illustrated as truncating the Firetrail anticline, an interpretation that is corroborated by drilling to the south (Lageson, 1984, Figure 10). This crosscutting relationship implies that the latest movement on the Absaroka thrust postdates at least early movement on the Firetrail thrust. Such a slowed progression of fault movement is often termed "out of sequence" thrusting. Hunter (1987) reported a similar local succession of thrusting for the Prospect thrust system in the vicinity of Battle Mountain (Figure 74) and Hoback Canyon (Day 8). In the northern half of the thrust belt such out of sequence thrusting is attributed to

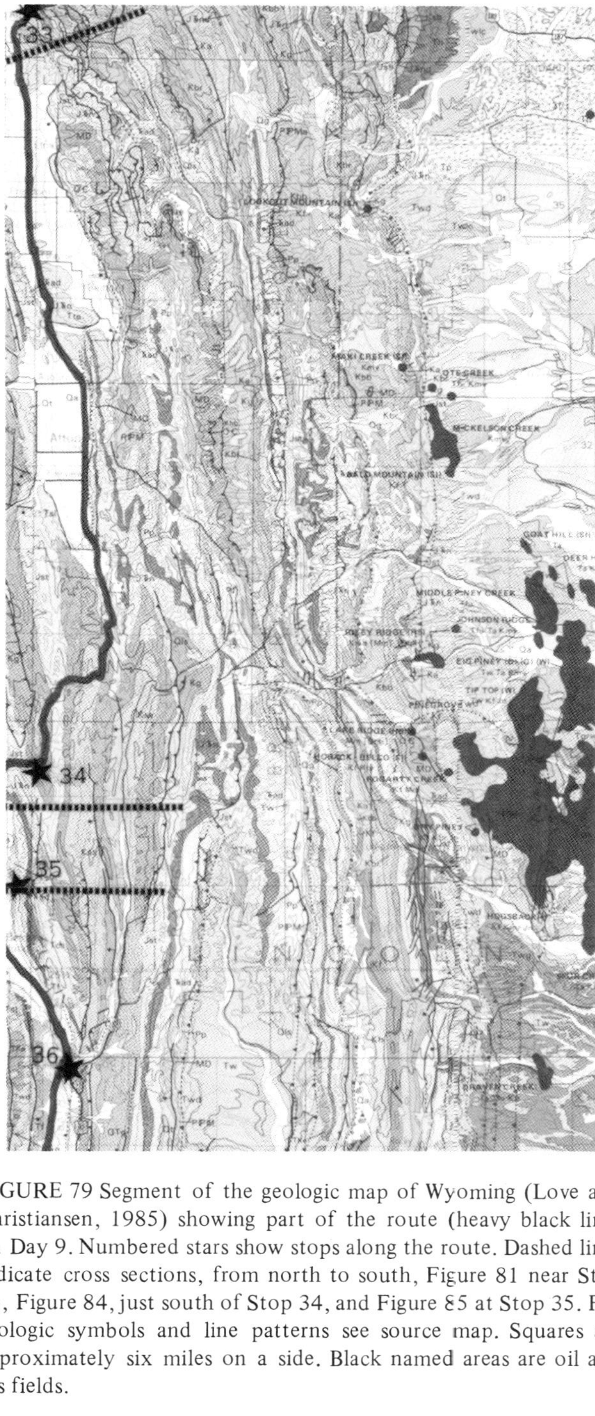

FIGURE 79 Segment of the geologic map of Wyoming (Love and Christiansen, 1985) showing part of the route (heavy black line) on Day 9. Numbered stars show stops along the route. Dashed lines indicate cross sections, from north to south, Figure 81 near Stop 33, Figure 84, just south of Stop 34, and Figure 85 at Stop 35. For geologic symbols and line patterns see source map. Squares are approximately six miles on a side. Black named areas are oil and gas fields.

the effects of buttressing of thrust movement by nascent uplift of foreland basement structures to the east.

The Alpine area also illustrates the common localization of Miocene to Holocene Basin and Range extensional faults along former thrust ramps. Here the Grand Valley normal fault is thought to be controlled by the Paleozoic to Cretaceous footwall ramp of the Absaroka thrust system (Royse, and others, 1975, Figure 6). Further steepening of the Absaroka ramp by the underlying Firetrail thrust (Figure 81) may have contributed to rotating the fault' plane to a mechanically favorable orientation for later extension.

The drive south from the Alpine area to Salt River Pass traverses a linked system of half grabens that form Grand and Star Valleys. The faceted mountain fronts of the Salt River Range to the east are the geomorphic expression of a series of normal faults that terminate and transfer along strike with an *en echelon* map pattern. Some of the movement is Pleistocene or Holocene. The clearest example of this kind of transfer is in The Narrows south of Thayne (Figure 79). Here, displacement on the Bedford normal fault (Blackstone and DeBruin, 1987) decreases markedly directly southeast of Thayne as the lower Star Valley normal fault gains displacement southwest of Thayne. This structure has been partly responsible for the preservation, in the intervening area, of soft Pliocene strata.

Several highway cuts south of Thayne expose red-brown to tan very soft claystones and sandstones that contain upper Pliocene fossils. These strata extend south into The Narrows where there is a unique occurrence of fossiliferous lignite in a coal prospect east of the Jurassic rocks and south of Willow Creek. These strata have not been studied but if the Pliocene age, suggested by the invertebrate fossils, is confirmed, this would be the only Pliocene lignite bed in the region.

On the east side of Star Valley, the Phosphoria Formation is exposed in a series of linear outcrops on the flanks of several faulted anticlines and synclines. The Meade Peak Member contains phosphorite beds and one vanadiferious mudstone (Figures 82, 83). This vanadiferous zone was sampled in 88 trenches and seven experimental mines by J.D. Love during 1942 and 1943 in Dry Creek Canyon southeast of Afton. In the region of Figure 82, the vanadiferous zone averages 4.15 feet in thickness. The richest three feet ranges in vanadium content from a minimum of 0.7% V_2O_5 in the northern part of the area to a maximum of 1.0% in the central part. It is estimated that in the area of Figure 82, there is 45 million tons of rock with an average grade of 0.9% V_2O_5 and average thickness of three feet above a depth of 500 feet (Love, 1961).

In addition to the vanadium, the black shales and mudstones in the Phosphoria Formation are also a veritable storehouse of other trace elements. Maximum values in this region are, in percent: Se 0.06, MoO_3 0.1, ZnO 1.3, TiO_2 1.0, NiO 0.3, Cr_2O_3 0.5. The black shales and mudstones in the Phosphoria elsewhere in the Thrust Belt also contain significant amounts of gold and silver. These have not been studied in as much detail as the vanadium, but values of as much as $4.50 per ton gold (at $400.00 per oz) and $22.40 per ton silver (at $10.00 per oz) have been determined (Love, 1984).

Stop 34

The Salt River Pass is an area of salt decollement and

GURE 80 Air oblique photograph, view north showing the structure on the north side of Snake River Canyon :st of Stop 32. The St. John thrust juxtaposes the Ordovician Bighorn Dolomite over Pennsylvanian and ssissippian strata in the footwall. The Cambrian outcrops at the left are in the hanging wall of the Ferry thrust. otograph by Bob Lynn, AMOCO, annotated by J.C. Coogan.

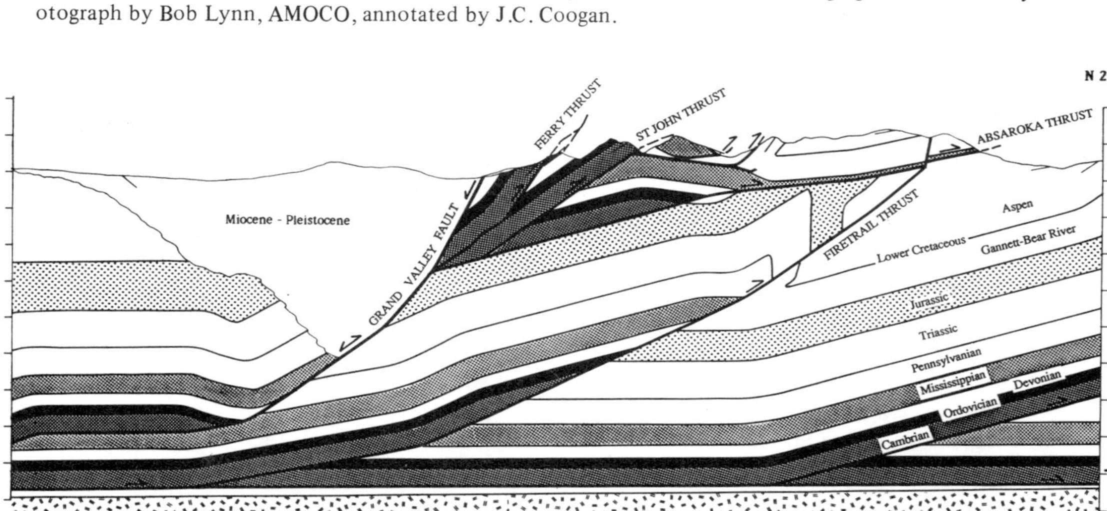

FIGURE 81 Cross section through the north side of Grays River Canyon. Figure 79 shows location just southeast of Stop 33. Generalized from Coogan, unpublished cross section.

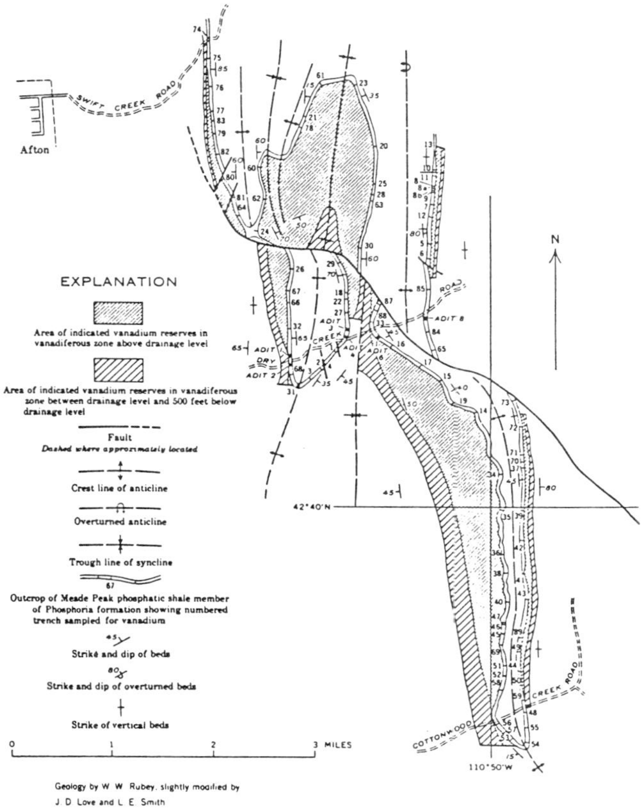

FIGURE 82 Index map of the Afton area showing outcrops of the Meade Peak Member of the Phosphoria Formation, the generalized structure, and location of 88 sampled trenches and seven adits in which vanadium ore was mined (Love, 1961).

the presence of an old molasse sequence. The route transects symmetrical folds in Jurassic and Lower Cretaceous strata in the footwall of the Crawford thrust. The folds have a horizontal amplitude of about three miles and a vertical amplitude of about 10,000 feet. The tightest folding is restricted to the redbeds in the Jurassic Preuss Formation and the younger units. In addition, the Preuss is the oldest unit carried by thrusts that cut this fold train (Porcupine Creek thrust and Muddy Ridge backthrust Figures 77, 84, and 85). A salt interval in the lowermost redbeds in the Preuss is the primary decollement for Crawford thrust shortening in this area. The salt manifests itself surficially by the emergence of numerous salt springs that are generally found in solution valleys that occupy the cores of anticlines such as Afton, and along the traces of major thrust faults, for example, along the Meade thrust to the northwest of the stop (Coogan and Yonkee, 1985). In the subsurface, the salt interval, where it has been drilled into, has been abnormally thickened in the Crawford thrust zone at depth. Figure 84 shows that more than 5,500 feet of salt was encountered in the core of the Afton anticline directly east of the Crawford ramp anticline. A normal salt thickness of 300 feet was encountered in an adjacent well 1.2 miles to the east. Stop 34 is in the core of the Afton anticline where the surface salt deposits can be seen. This is the site of abandoned pioneer salt works. The annual production of salt from this general area in 1877 was 200,000 pounds

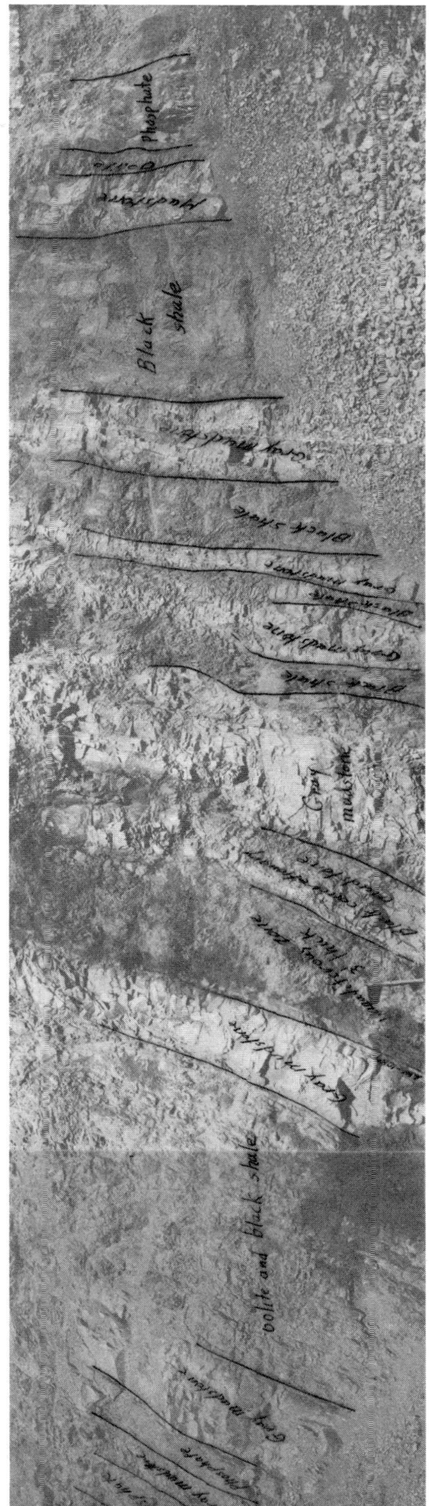

FIGURE 83 Trench 74 on Swift Creek (see location on Figure 82) exposing the Meade Peak Member of the Phosphoria Formation. Oldest beds are at right; base of Rex Chert Member is at extreme left. Shovel on vanadiferous zone indicates scale. Photograph by J.D. Love (1967).

(Peale, 1879, p. 645)! Indians likewise quarried the salt here. Also, at Stop 34, conglomerates in the Lower Cretaceous Gannett Group can be seen. This is the oldest molasse deposit associated with thrusting in the Thrust Belt.

Stop 35

The Sublette anticline is the main surface expression of the northern Crawford thrust. Well control on the forelimb and backlimb of the anticline (Figures 84, 85) demonstrates that it is the result of a hanging wall ramp in the Crawford thrust plane from the Paleozoic level to the Jurassic Preuss decollement. The present structural elevation of Sublette anticline is the composite result of uplift above the Cokeville and Afton anticlines, as well as uplift above the Crawford ramp itself. The Afton anticline plunges south from Stop 34 into the line of section in Figure 84. The Cokeville anticline is viewed at Stop 36, where it plunges north into the line of section. The route

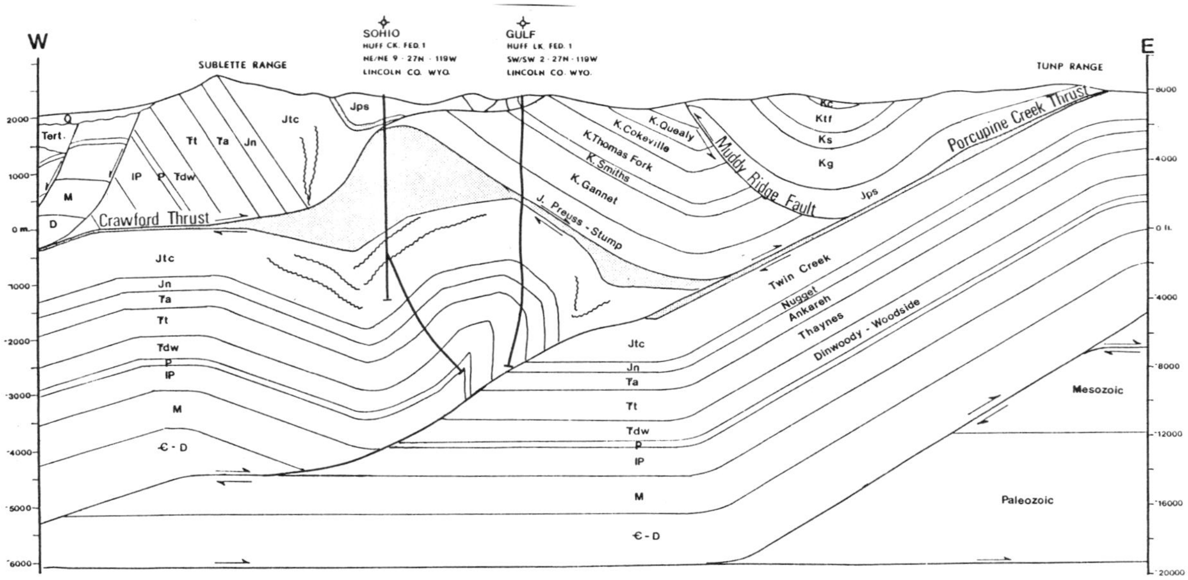

FIGURE 84 Cross section illustrating the Preuss salt decollement zone (stippled) east of the Sublette anticline. From Coogan and Yonkee (1985).

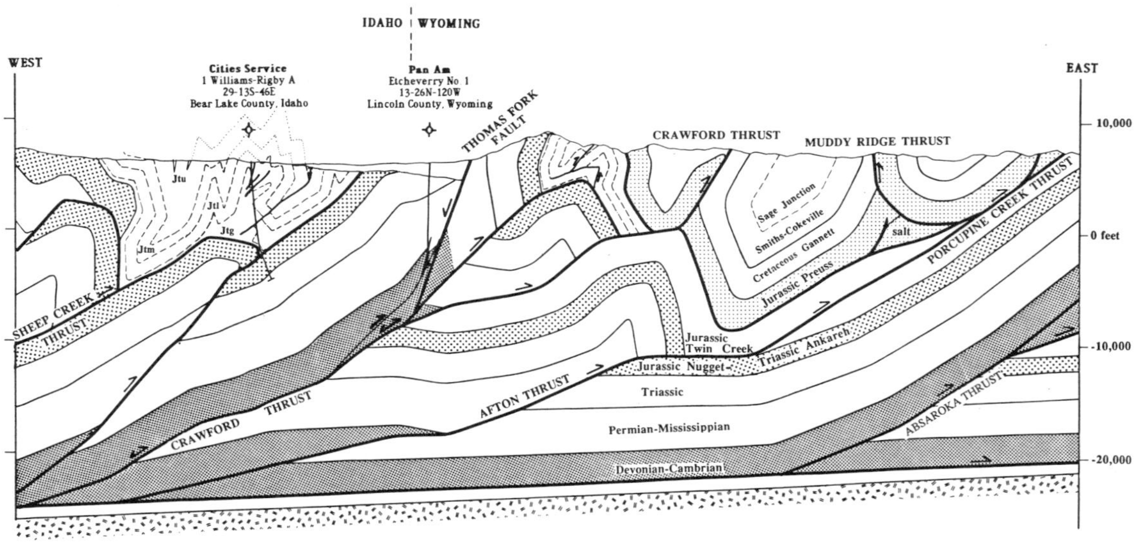

FIGURE 85 Cross section of the Crawford thrust sheet along the south side of Raymond Canyon and areas to the east and west. Indicated are: Jtg – Gypsum Spring Member, Jtl – lower members (Sliderock and Rich), Jtm – middle members (Boundary Ridge and Watton Canyon), and Jtu – upper members (Leeds Creek and Giraffe Creek) of the Twin Creek Limestone. Generalized from J.C. Coogan, unpublished cross section.

into Thomas Fork valley transects the north plunging nose of the Sublette anticline in Salt River Canyon. Here, the Jurassic Twin Creek Limestone is folded into a box fold where the steep limbs and flat crest can be seen on the north canyon wall.

Along the route south through Thomas Fork valley, the Sublette anticline can be seen at successively deeper stratigraphic levels to the east, until Pennsylvanian strata are exposed in the core near Raymond Canyon. The faceted mountain front marks the approximate location of the Thomas Fork normal fault. As with the Grand Valley fault and the Absaroka thrust ramp, the Thomas Fork fault is localized on the oversteepened Crawford thrust ramp. The approximate fault location and throw are conveniently constrained by the well control directly west of the mountain front. Throw on the fault is approximately 3,500-4,000 feet along the line of section.

A spectacular decollement fold train on the west side of Thomas Fork valley can be seen from the turnoff to Raymond Canyon. It will be discussed at this stop.

Raymond Canyon presents a well exposed cross section through the overturned limb of a tightened ramp anticline (Figure 85). The canyon cuts from the Pennsylvanian Wells Formation through the Jurassic Twin Creek Limestone and permits identification of the different structural styles encountered in the various lithologies. The entrance to the canyon is marked by vertical walls of highly fractured quartzite in the Wells, but is most notable for the exposures of the Rex Chert Member of the Permian Phosphoria Formation which Mansfield (1927) described as resembling "crushed organ pipes". In 1943, two adits were driven along the vanadiferous zone in the Meade Peak Member of the Phosphoria Formation on opposite sides of this canyon, the south one by the U.S. Bureau of Mines and the north one by Homestake Mining Company. During the 1960s, the north adit was driven deeper by another mining company, looking for higher values of vanadium and other trace elements below the zone of weathering. The results have not been released.

Higher in the Raymond Canyon section, small scale kinking and "out of the syncline" folding and thrusting are present in the shaly intervals of Triassic strata. The Twin Creek Limestone displays well developed refracted cleavage and small scale thrusting at the east end of the canyon.

Stop 36

The Sublette anticline plunges south from Raymond Canyon toward Cokeville. Just northeast of the town, the frontal structure of the Crawford thrust is the Cokeville anticline (Figure 86), which documents a local hanging wall ramp from the base of the Wells Formation onto the Preuss salt decollement (Coogan and Yonkee, 1985). The isoclinal core of the Cokeville anticline is representative of many of the subsurface anticlines encountered in oil and gas exploration of the Crawford thrust sheet (for example, the Afton anticline, Figure 84 and the Hogsback Ridge field; Valenti, 1982). In many of these wells, anhydrite was encountered in the lower part of the Wells Formation, which forms the isoclinal cores of the structures. Flowage of anhydrite may explain the hinge collapse seen in the Cokeville anticline. An old phosphate adit was driven along the Meade Peak Member of

FIGURE 86 Cokeville Butte, a mile north of Cokeville. This is a part of the Cokeville anticline on the hanging wall ramp of the Crawford thrust. Indicated are: IPw – Wells Formation, Pp – Phosphoria Formation, Ŧrd – Dinwoody Formation, Ŧrw – Woodside Formation, Ŧrt – Thaynes Formation. Photograph by J.H. Rathbone.

the Phosphoria Formation along the east flank of the anticline.

The latitude of Cokeville marks a fundamental change in the geomorphology and surface geology of the Wyoming-Idaho-Utah Thrust Belt. It also marks a change in distribution of economic resources. Is this coincidence or not? The future will tell us. North of this latitude there are huge oil and gas fields near the eastern margin of the Thrust Belt, in fields like LaBarge, Big Piney, Dry Piney, etc. (Figure 79). In contrast, huge oil and gas fields are within the Thrust Belt (Figure 87), south of this latitude. Similarly, coal resources are abundant to the south, on the east edge of the Thrust Belt, but much smaller to the north.

South of Cokeville, the mountain ranges of folded and faulted Paleozoic rocks that are so prominent and conspicuous farther north are replaced by broad plateaus of gently warped uppermost Cretaceous through lower Eocene strata. The surface area between these two stratigraphic packages exhibits the principal angular uncon-

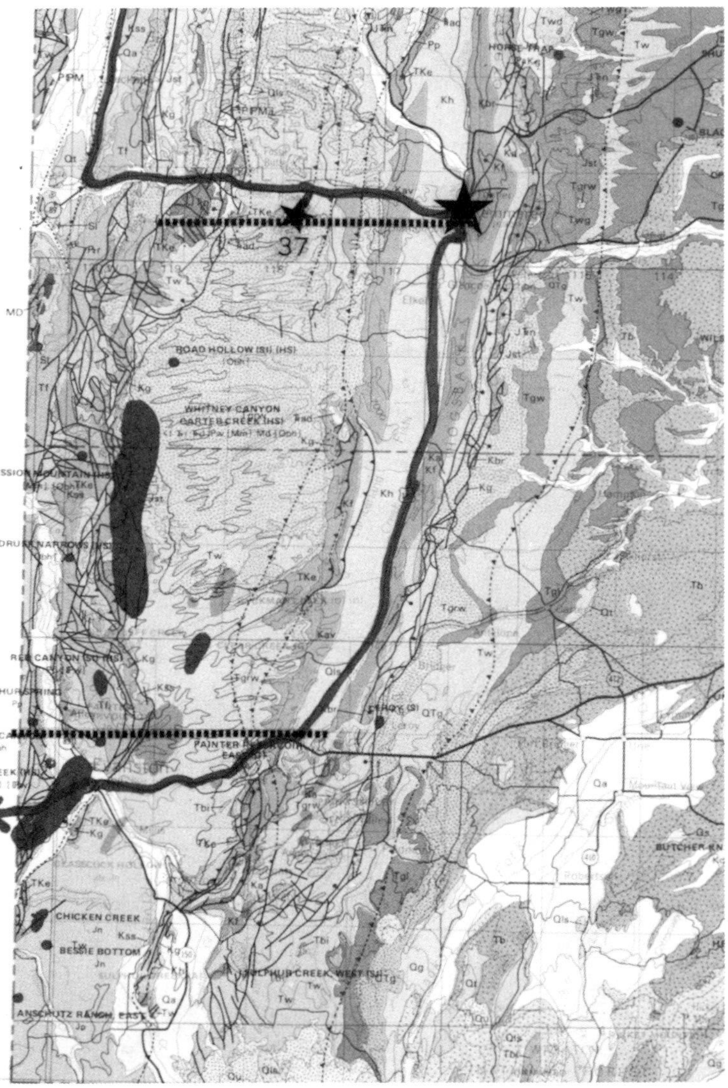

FIGURE 87 Segment of the geologic map of Wyoming (Love and Christiansen, 1985) showing parts of the route for Day 9 and Day 10 (heavy black line). The large star is at the overnight stop in Kemmerer. The smaller numbered star is the Fossil Basin stop. The dashed lines show the positions of cross sections, Figure 88 near Kemmerer and Figure 95 near Evanston. Black named areas are oil and gas fields. For geologic symbols and line patterns see source map. Squares are approximately six miles on a side.

formity in the region. This unconformity developed diachronously across the area, depending on the position of the underlying thrust structures that controlled the loci of erosion and deposition from latest Cretaceous through early Eocene (Figure 77; Wiltschko and Dorr, 1983). Late phase movement on the Absaroka thrust occurred during the latest Cretaceous (Maestrichtian) (Royse, and others, 1975), with continued minor uplift through the early Eocene that was localized at both the thrust front and above the Absaroka footwall ramp and associated thrust imbricates (Hurst and Steidtmann, 1986). Late reactivation and uplift also took place along the Crawford thrust front during the early Eocene. The depositional basins between these uplifted areas were continuously translated eastward above the basal decollement, first by the Absaroka thrust, and later by the Hogsback thrust. Basins that are formed on the hanging wall above an active decollement are commonly called "piggyback" basins (Ori and Friend, 1984).

The drive between Cokeville and Fossil Butte National Monument (Figure 87) skirts along the eastern edge of one piggyback basin, preserved in Bear Lake Plateau, and transects the second and better known Fossil Basin. From Cokeville to Sage Junction, the eastern edge of Bear Lake Plateau is seen in the dissected red-weathering hillsides directly to the west. The red strata are chiefly lower Eocene debris- and mud flow deposits in the Wasatch Formation. The debris flows thicken eastward toward the Crawford thrust front and contain many angular clasts of Triassic and Jurassic rocks common to this front. These chaotic deposits grade westward into braided stream conglomerates that have a southerly axial paleocurrent direction (Coogan, unpub. data). Lacustrine limestones cap the fluvial sequence farther west. The general early Eocene depositional pattern of Bear Lake Plateau is one of poorly sorted coarse clastics supplied to alluvial fans on the tectonically active basin margin. These intertongue with and grade upward into better stratified and generally finer grained fluvial and lacustrine deposits toward the basin center. The Fossil Basin presents a similar depositional pattern that is better exposed and more easily transected.

The prominent range visible south-southwest of Sage Junction is the Crawford Range (Figure 87), which consists of Paleozoic carbonate rocks that define the leading edge of the Crawford thrust plate. The Crawford Range marks the western edge of the lower Eocene Fossil Basin south of Sage Junction. This margin has subsequently been modified by Pliocene (?) to Holocene Basin and Range normal faults.

About three miles west of Sage Junction is a huge open-pit phosphate mine in the Phosphoria Formation. It was opened in 1947 and has produced several million tons of phosphate fertilizer from a shallow syncline in this formation. At present, the mine is inactive.

Eastward from Sage Junction, the geology along the route all the way to Kemmerer has been mapped in detail by Rubey, and others (1975). The western margin of

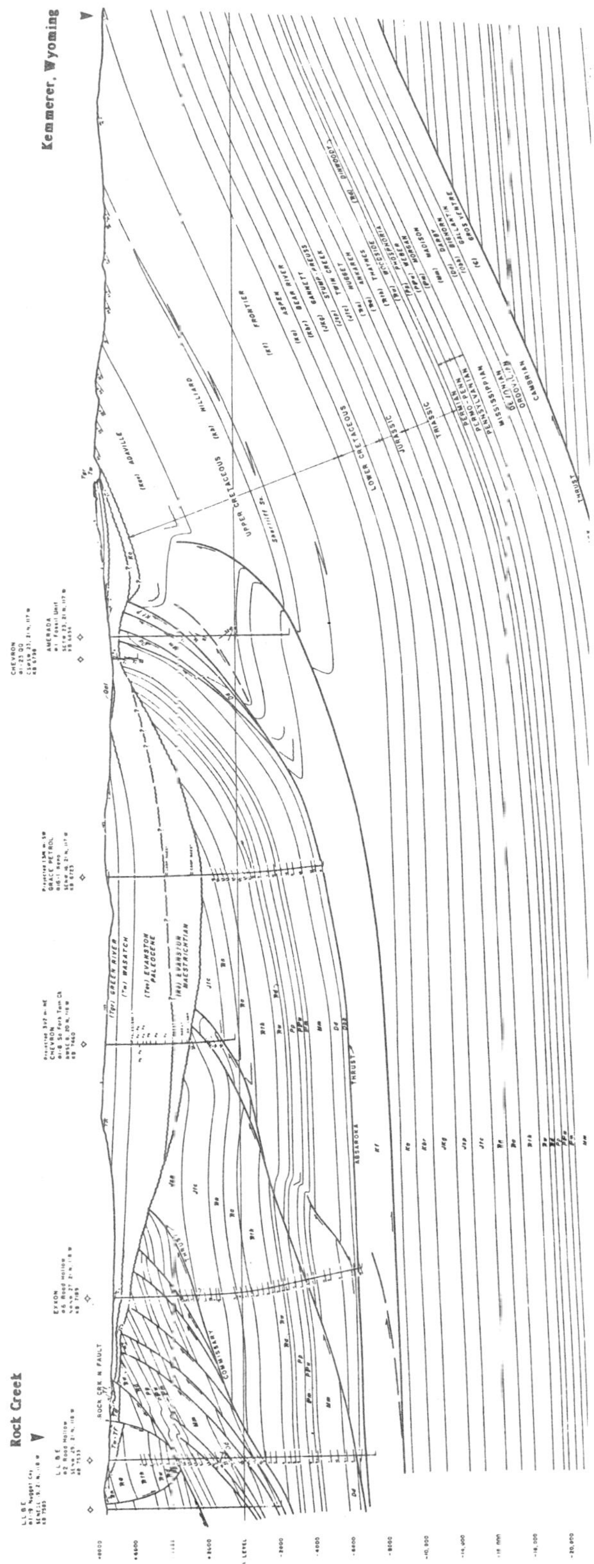

FIGURE 88 Cross section from Rock Creek to Kemmerer. For location, see Figure 87 near Stop 37. From Lamerson (1982).

Fossil Basin is defined by homoclinally west-dipping Mesozoic strata above the footwall ramp of the Absaroka thrust (Lamerson, 1982). A series of young normal faults can be recognized along the north side of the highway where hangingwall Tertiary strata are juxtaposed against footwall Mesozoic rocks. The Mesozoic exposures include the type localities for both the Jurassic Twin Creek Limestone and Jurassic or Triassic Nugget Sandstone, which are well exposed on the north side of the highway. The Nugget Sandstone is a prolific oil reservoir in the Ryckman Creek field 26 miles to the southeast. In the Lake Alice area, 25 miles northeast of Cokeville, and elsewhere in the Thrust Belt for 100 miles to the north, the Nugget contains sporadic, but interesting, occurrences of copper, silver, and zinc (Love, and Antweiler, 1973; Loose, and Boberg, 1987). At Lake Alice there was extensive mining for copper in the early part of the 20th century.

West of Fossil Basin, the Rock Creek normal fault is well expressed by a long, narrow strike valley. The conspicuous flatirons seen at the mouth of the valley are part of an overturned section of the Triassic Thaynes Formation in the footwall of the fault. The small hills at the base of the flatirons are early Eocene Wasatch and Green River Formations (these tongues of the Green River are considered to be late early Eocene (Grande, 1980) that dip as much as 45° east into the fault. This is typical for listric faults that shallow with depth. A drill hole through the fault plane (Figure 88) documents this geometry, with the Rock Creek fault soling into an earlier thrust imbricate zone above the Absaroka thrust plane.

The early Eocene stratigraphy of Fossil Basin is well displayed along the highway to Fossil Basin National Monument (Figure 89). Lower Eocene strata are divided between the Wasatch Formation fluvial sequence and the intertonguing Green River Formation lacustrine sequence (Oriel and Tracey, 1970; Grande, 1980). The lower

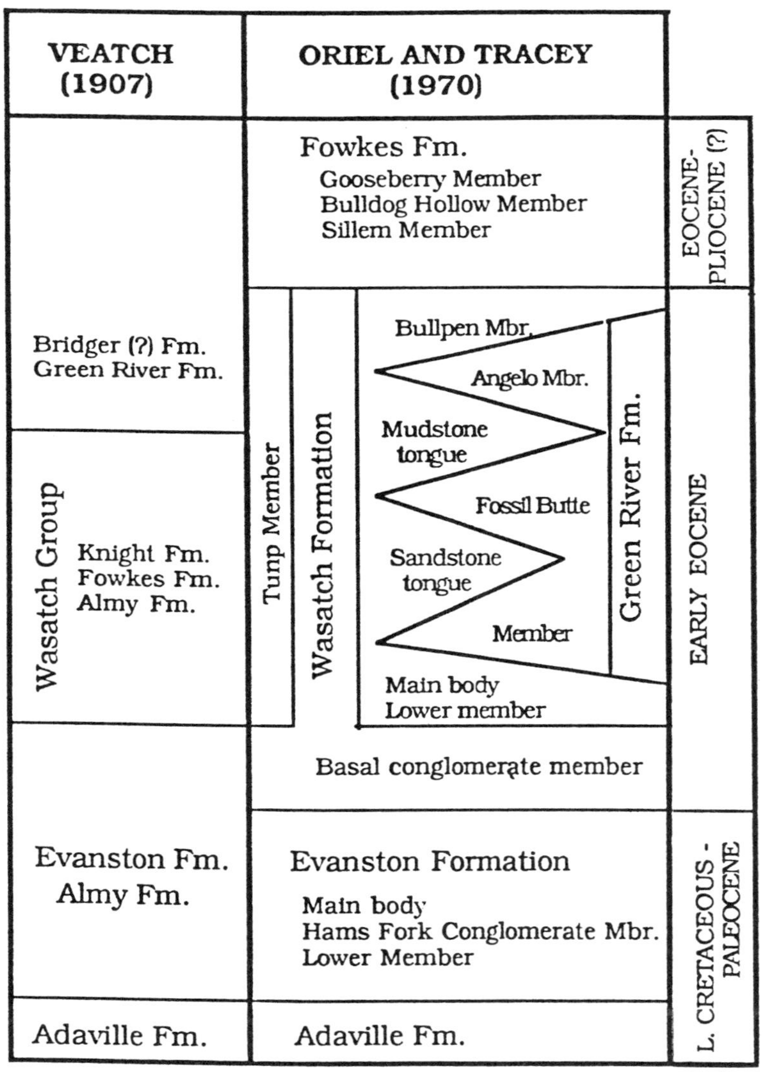

FIGURE 89 Correlation diagram of latest Cretaceous through Eocene strata of the Fossil Basin. From Oriel and Tracey (1970).

member of the Wasatch Formation consists of gray claystone, variegated mudstone, and tan to red sandstone and conglomerate. The sole outcrop of the lower member along the trip route is located 1.5 miles east of Rock Creek, directly past the railroad bridge, where sandstone ribs dip 10°-15° into the basin center. The stratigraphy of the basin center can be seen by looking northwest toward Fossil Butte. The main body of the Wasatch Formation consists of red, purple, and yellow mudstone that is exposed in the foreground and on the lower slopes of Fossil Bute. A prominent vegetation band marks the contact between the Wasatch and Green River Formations. The Fossil Butte Member of the Green River Formation (200-280 feet thick) forms the tan cliffs directly above this contact, and is composed of mudstone, calcareous sandstone, siltstone, fissile shale, limestone, and oil shale. The main fossil fish bed is located 50 feet below the top of the member. The Angelo Member of the Green River Formation caps Fossil Butte and consists of light colored limestone, marlstone, and mudstone.

Stop 37

This stop is off the highway and up the hill to the south at the Ulrich fossil fish quarry. Because of the oil shale, the fish fossils, trona (sodium sesquicarbonate, $Na_3(CO_3)(HCO_3) \cdot 2H_2O$) deposits and uranium, the Green River Formation has been studied for more than 100 years. The first report of fossil fish was by Dr. John Evans in 1856 (Grande, 1980). The Territorial Surveys also collected fossil fish from many sites in southwestern Wyoming and most of these, including some from near Stop 37, were described in a monumental study by Cope (1884). These studies have been brought up to date by Grande (1980). Private collectors began selling specimens before 1900 and have continued to the present time. Carl and Shirley Ulrich opened their quarry at Stop 37 in 1947 and have operated it continuously since then (Figures 90; 91).

The lakes in which the Green River Formation was deposited represent one of the most unique events in geologic history. In addition to their size (Figure 92), longevity (from 58 to 38 million years ago; Grande, 1980), and unique history of sedimentation (Surdam and Stanley, 1980), they also contain enormous deposits of oil shale (Bradley, 1964), trona (Univ. Wyo. Contrib. to Geology, Trona Issue, 1971), and lesser amounts of uranium (Love, 1964) and at least 70 exotic mineral assemblages (Milton, 1971).

Around the western, northern, and eastern peripheries of Fossil Basin, the coarse grained debris flow deposits of the Tunp Member of the Wasatch Formation (Figure

FIGURE 90 Ulrich fossil fish quarry at Stop 37, in the Fossil Butte Member of the Green River Formation (late early Eocene). Left to right, Shirley and Carl Ulrich.

FIGURE 91 Fossil fish *Diplomystus dentatus,* about 10 inches long, one of the more common fish in the Ulrich quarry at Stop 37.

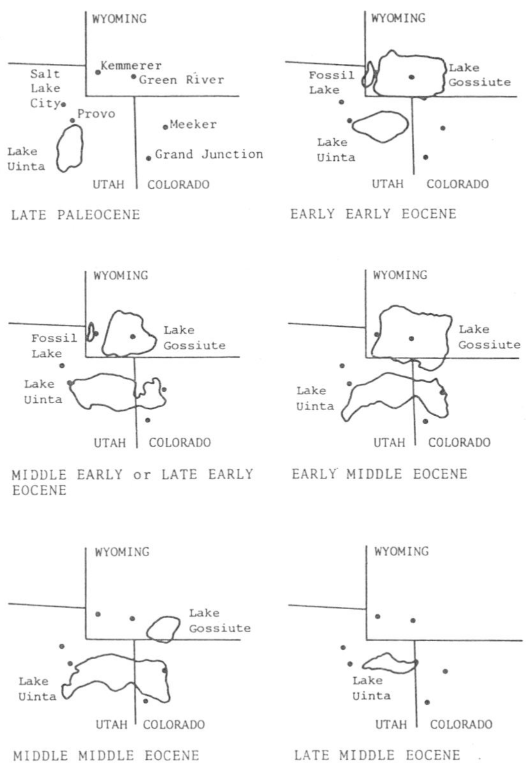

FIGURE 92 Approximate locations and areas of the three Green River lakes at six different intervals of time during their history. From Grande, 1980.

89) intertongue with the fine-grained deposits of the basin center (Oriel and Tracey, 1970). The Tunp Member crops out north of the field trip route, where its source areas are correlated to areas of thrust uplift during deposition (Hurst and Steidtmann, 1986). Figure 93 shows the intertonguing of the Tunp with more basinal units adjacent to three source areas: the Boulder Ridge section lies directly east of the Crawford thrust front; the Dempsey Ridge section is east of the Absaroka ramp imbricates; and the Commissary Ridge section lies west of the Absaroka thrust front on the eastern margin of Fossil Basin.

Uplift of the eastern basin margin is also documented in outcrop and subcrop by the Paleocene-Maestrichtian Evanston Formation. The Evanston overlies, with an angular unconformity, older Mesozoic and Paleozoic strata in outcrops along the eastern basin margin as well as in the basin center where the unconformity is documented by well control (Figure 88). A second angular unconformity between the Maestrichtian part of the Evanston (lower member and Hams Fork Conglomerate Member) and the Paleocene main body of the Evanston is above the Absaroka thrust front. This contact becomes disconformable toward the basin center. The basin margin unconformity is directly related to slip along the Absaroka thrust plane. The Maestrichtian part of the Evanston is cut by the Absaroka thrust in the subsurface during a late, minor (5,000 feet displacement) phase

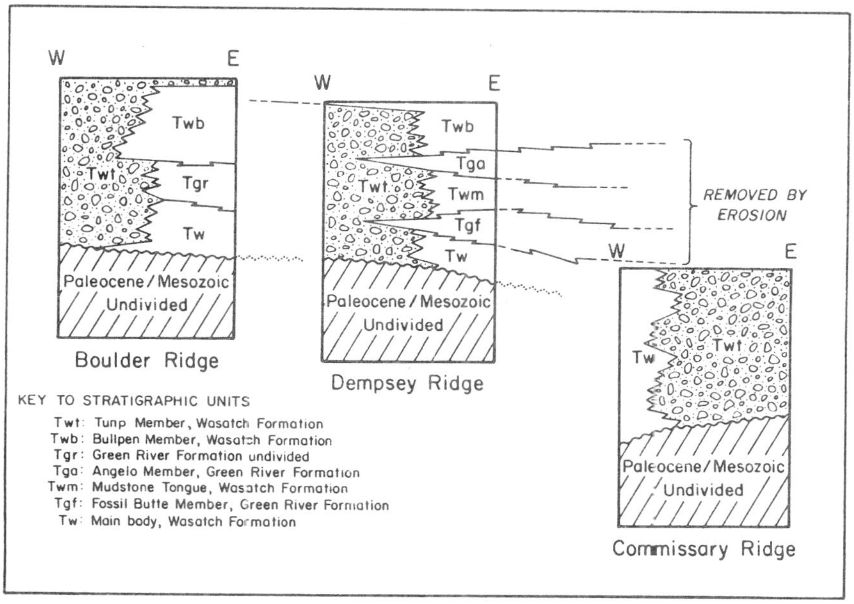

FIGURE 93 Diagram showing intertonguing relations between members of the Wasatch and Green River Formations in the northern part of Fossil Basin. From Hurst and Steidtmann (1986).

of slip along the Absaroka thrust (Figure 88). The Maestrichtian section and the thrust tip are angularly overlapped by the Paleocene part of the Evanston, which is gently folded above the thrust. In summary, the Evanston unconformity effectively dates a final, minor phase of slip across the Absaroka thrust at this location as occurring near the Cretaceous-Tertiary boundary. Two distinct earlier phases of major Absaroka slip were recorded by Royse, and others, (1975) based on similar stratigraphic bracketing in the Little Muddy Creek area 13 miles to the south. A mid-Santonian synorogenic conglomerate derived from a western Absaroka splay documents early slip on the Absaroka thrust, and an eastern splay that is overlapped by the Maestrichtian part of the Evanston Formation records the latest major slip of the Absaroka (Figure 94).

The eastern margin of Fossil Basin also displays a local angular unconformity between the lower Eocene Wasatch Formation and the underlying Evanston Formation at Trail Creek, 12 miles north of the field trip route (Figure 95). Here, the Tunp Member of the Wasatch Formation dips gently into the Fossil Basin above the Maestrichtian part of the Evanston Formation that dips as much as 60° west. The Tunp also angularly overlies the lower member of the Wasatch directly to the west. In total, these relationships document the sequential development of the uplift at the Absaroka thrust front that was the source of the coarse clastic deposits of the Tunp Member along the eastern margin of the Fossil Basin. Since this uplift postdates the late, minor slip along the Absaroka thrust plane discussed earlier (Figure 88), Hurst and Steidtmann (1986) have proposed that early Eocene uplift of the

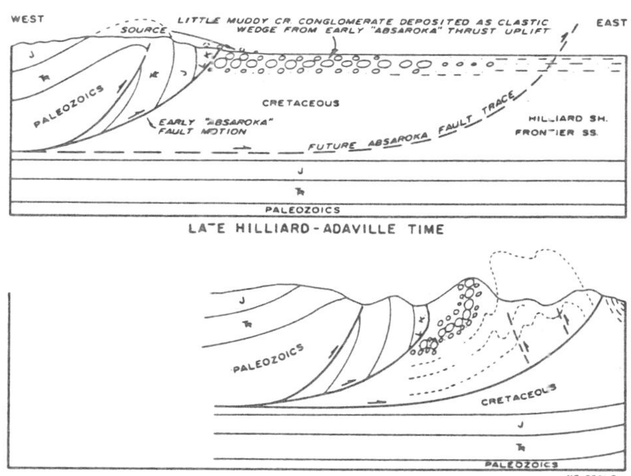

FIGURE 94 Diagram illustrating sequential development of the Absaroka thrust system in the Little Muddy Creek area. From Royse, and others (1975).

Absaroka front was the result of Absaroka footwall uplift due to ramping of the Hogsback thrust beneath it.

Boulder beds of the Hams Fork Conglomerate Member of the Evanston are accessible along the north side of the highway route to Kemmerer. The clast provenance of this conglomerate is of particular interest. Although the Hams Fork does contain some locally derived Mesozoic clasts from the Absaroka thrust front (Lamerson, 1982), it also contains distinctive arkosic quartzite clasts that have a unique provenance in the Lower Cambrian and Proterozoic of the Willard and Paris thrust plates located 40

FIGURE 95 Air oblique photograph illustrating structural and stratigraphic relationships along the eastern margin of the Fossil Basin at Trail Creek, 12 miles north of the field trip route. Indicated are: TRt – Thaynes Limestone, TRa – redbeds of the Ankareh Formation, JTRn – Nugget Sandstone, Jt – Twin Creek Limestone, Kel – lower member of the Evanston Formation, Keh – Hams Fork Conglomerate Member of the Evanston Formation, Tw – Lower member and main body of Wasatch Formation, Twt – Tunp Member of Wasatch Formation. Note the angular unconformity between the Tunp Member and underlying units. Photograph by Bob Lynn, AMOCO. Annotated by J.C. Coogan.

miles to the west. The presence of these clasts demonstrates that the Fossil Basin was poorly developed in Maestrichtian time, allowing boulder-size clasts to be transported from the west. The provenance of the quartzites also demonstrates that the hinterland source area was uplifted concurrent with slip on eastern thrusts (Schmitt, 1987).

The drive between Fossil Basin and Kemmerer transects the west-dipping homocline in Upper Cretaceous strata that are folded above the Hogsback footwall ramp (Figure 88). The Elkol coal mine directly south of the highway is in the fluvio-deltaic Adaville Formation, which was deposited concurrently with early Absaroka thrust motion ot the west. The town of Kemmerer straddles the contact between the Upper Cretaceous Hilliard Shale on the west and the sandstone hogbacks of the Frontier Formation on the east.

Kemmerer, Wyoming, Overnight Stop

The town of Kemmerer and adjacent villages such as Frontier and Diamondville, were established because of coal. It has been mined continuously since 1874, chiefly in the Upper Cretaceous Frontier and Adaville Formations. A total of about 145 million tons of bituminous and subbituminous coal has been mined from subsurface and strip mines. Two strip mines west of Kemmerer currently produce about 5 million tons of subbituminous coal per year from as many as 11 seams having a cumulative thickness of 300 feet. The thickest individual coal seams are 20, 50, and 100 feet. The Pittsburg and Midway Coal Mining Company mine is reportedly the largest (in area, not in production) open pit mine in the United States (R.W. Jones, written commun.). The coal beds dip west 17°-23° and a total depth of the main pit is 700 feet. Most of the coal is burned in steam-electric generating plants.

The route south from Kemmerer to Interstate 80 follows a strike valley in the Upper Cretaceous Hilliard Shale. This shale is of considerable interest because it is the eastern marine equivalent of the thick molasse sequence of the Echo Canyon Congolmerate that will be seen at Stop 38. Today, this is only 40 miles away but back in Laramide time it was much, much farther away. How much farther, we don't know. The hogbacks to the east of the route are sandstones in the Frontier Formation, and they mark the approximate location of the

Absaroka thrust front beneath the cover of Tertiary rocks in the Fossil Basin.

There is not much to see along the first 35 miles of this day's trip so here is an opportunity to reflect on the recent advances in geologic knowledge relating to the segment of the Thrust Belt directly to the west of this part of the route. Fourteen years ago (1975), there were no oil and gas fields to the west, yet now there are many small and large ones (Figure 87), including one "world class" field (Whitney Canyon-Carter Creek), and possibly a second one across the State line in Utah. Volumes have been written on the geology, geophysics, and oil and gas exploration and production during this 14-year period (for example, Wyoming Geological Association volumes on the Thrust Belt, 1977, 61 articles, 787 pages; 1987, 34 articles, 403 pages; Petroleum Information, 1981, regional overview, description of 17 fields, 251 pages).

So – why did it take the petroleum industry until 1975 to recognize the potential of this very accessible area in its own back yard? The pioneer geologist, A.C. Veatch, with two geological assistants, spent 72 days in 1905 mapping this entire productive segment of the Thrust Belt, 1,700 square miles, making his own topographic and base map, without benefit of motor vehicles, air photos, or geophysical data. By 1907, he had published a 178-page classic U.S. Geological Survey Professional Paper 56 with not only a colored geologic map but a second one in which the Tertiary cover had been stripped away. This report is titled: "Geography and geology of a portion of southwestern Wyoming with special reference to coal and oil". By means of many cross sections, he discussed the concept of thrust-faulted anticlines – almost heresy 70 years ago. The oil springs and the shallow wells (all but three under 2,000 feet; most under 1,000 feet deep), in the general area of today's fields were described in detail (pages 139-162). He concluded that the oil possibilities were excellent.

What happened in the next 70 years? A few dry holes, some geophysical work, but little inclination to risk much money in a serious exploration program. In the light of this contemporary example of myopia, one can wonder how many other bonanza areas remain undiscovered because they do not fit the current ruling theories. It is hoped that this trip will demonstrate that there are still many new frontiers of knowledge, many new and old concepts that need to be examined critically, and many innovative avenues of research that the modern geologist should not be afraid to explore. As Veatch showed by example, perceptive field geology has no substitute, and this is true even more so today, even in the era of computers and "black boxes".

The east-west-trending Uinta Mountains are visible to the south (Figures 97, 98; just south of the area of Figure 87). They mark the southern boundary of the Wyoming salient of the Cordilleran Thrust Belt. This trip turns west about 12 miles east of the town of Evanston, Wyoming. Figure 96 shows a cross section of the structure along this segment of the trip, and provides an interpretation of the

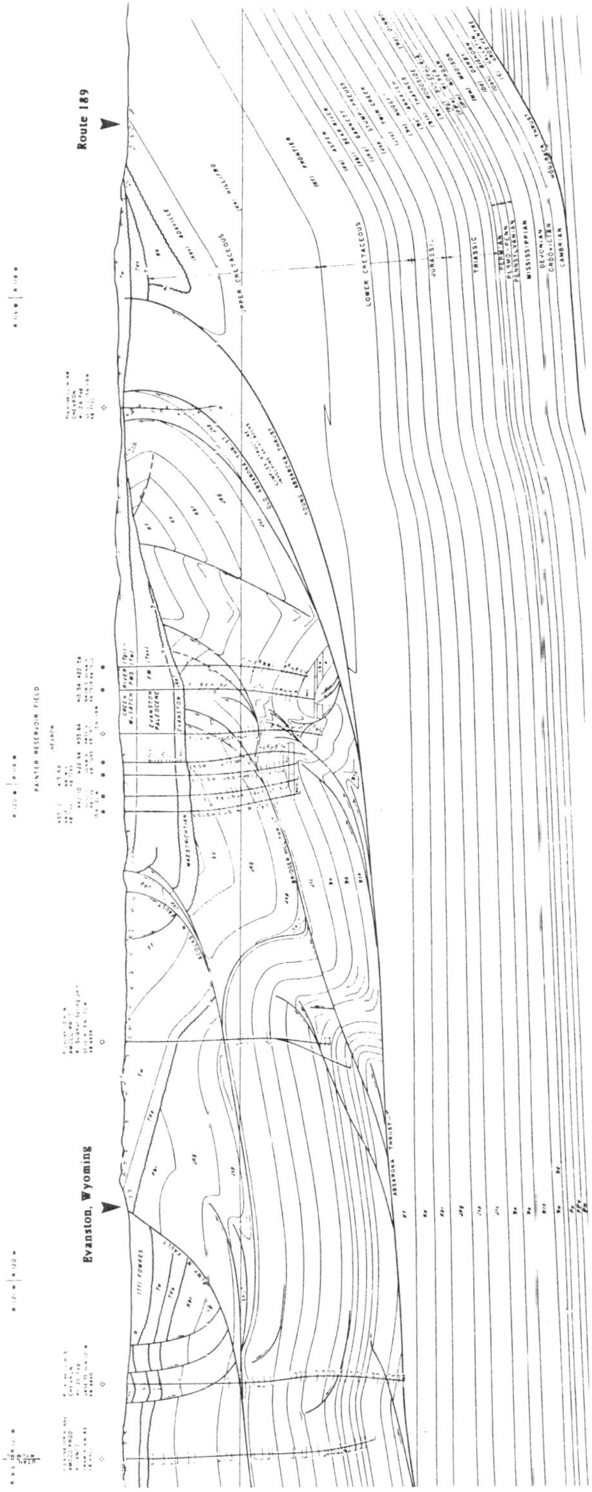

FIGURE 96 Cross section of the structure along the route between the junction of U.S. Highway 189 and Interstate 80 to, and west of, Evanston. From Lamerson (1982).

structural style of the oil and gas fields within the southern part of the Absaroka thrust sheet. The open fold geometries that form structural traps in this area contrast with the highly imbricated structural style of the northern part of the Absaroka sheet seen earlier at Alpine (Figures 79 and 81). The hangingwall anticlinal traps are supplied with hydrocarbons from footwall Cretaceous marine shales migrating across the Absaroka thrust plane (Warner, 1982). The principle oil and gas production in the Absaroka thrust is from two distinct trends of hangingwall ramp anticlines. The eastern trend represented in Figure 96 by Painter Reservoir field is the result of a hanging wall ramp in the Triassic Thaynes through Jurassic-Triassic Nugget Sandstone which generally produced low-sulfur gas, oil, and condensate. In contrast, the western trend, represented by the Sulphur Spring well (Figure 96), is the result of a hangingwall ramp through the Paleozoic section. The western trend produces from Paleozoic reservoirs, principally from the porous dolomites of the Mississippian Mission Canyon Limestone. The western trend production is limited to high sulfur gas and condensate, with sulfur originating, chiefly, in evaporites within the Mission Canyon. The difference in hydrocarbon maturity from east to west is presumably the result of slight differences in burial history (Warner, 1982). In both cases, the migration path is thought to be through the hanging wall cutoffs of the reservoir formations on the eastern limb of the anticlines (Lamerson, 1982). The Preuss salt decollement seen earlier at Salt River Pass (Stop 34) is also an important feature in the southern part of the Thrust Belt, accommodating disharmonic folding above the producing structures.

The Preuss salt decollement also controlled the late thrusting and later Basin and Range extension seen near Evanston (Figure 96). The Medicine Butte thrust cuts the Evanston and Wasatch Formations there and represents thrusting well after the main movement on the Absaroka thrust. The salt decollement of the Medicine Butte thrust also served to localize later extensional faults such as the Acocks and Almy normal faults. Once again, the listric shape of the normal faults is confirmed by drilling (Figure 96) and the characteristic rotation of Tertiary strata into the fault plane is well displayed in the 20°-45° east dips of the Evanston through Fowkes Formations in the Acocks fault hangingwall on the north side of Evanston.

Red and brown conglomerates and mudstones of the Wasatch Formation are continuously exposed for 25 miles west of Evanston along Interstate 80 (Figure 97). At approximately 25 miles west, an important angular unconformity between the lower Eocene Wasatch Formation and the underlying Coniacian-Santonian Echo Canyon Conglomerate is exposed on the south side of the highway. The Echo Canyon is the uppermost horizon on the east flank of the Stevenson syncline, a fold formed on the backlimb of the Absaroka thrust above the Absaroka footwall ramp (Figure 97). Farther west along Echo Canyon, more than 1,700 feet of dull red coarse boulder and cobble alluvial fan conglomerate is exposed along the canyon walls. Royse, and others, (1975) first proposed that the Echo Canyon Conglomerate was derived from uplift of the Crawford thrust to the northwest. More recently, however, DeCelles (1988) has determined that the conglomerate clasts have a mixed provenance from both the Proterozoic-Paleozoic rocks of the Willard thrust hanging wall, and the Paleozoic-Mesozoic section above the basement-cored anticlinorium east of Ogden and Salt Lake City (Figures 75, 97, 98). The Wasatch basement uplift is recognized as a ramp anticline that has been translated across the basement decollement that is shared by the Absaroka and Hogsback thrusts to the east (Royse, and others, 1975).

Essentially, the Wasatch uplift represents basement shortening that corresponds in timing and magnitude to the "thin skinned" shortening expressed by the thrusts to the east. Coogan (1987) noted that the Crawford thrust also shares the basement decollement with the basement uplift. Thus, the initial basement uplift recorded by the Echo Canyon provenance may indirectly correlate with movement on the Crawford thrust as suggested by Royse, and others (1975).

Stop 38

This stop is located so that the angular unconformity between the Wasatch Formation and the underlying Echo Canyon Conglomerate can be seen. In addition, there will be an opportunity to take a close look at the sedimentary features displayed in the canyon walls.

At Echo Junction, the route turns northwestward on U.S. Highway 84 for about 10 miles to the Devils Slide area.

Stop 39

The Devils Slide is a series of topographic ridges and swales within the Jurassic Twin Creek Limestone on the eastern flank of the Wasatch basement-cored uplift. This is the first time the route has visited a basement-cored uplift since Stop 30 on Day 8. Proximal, poorly sorted boulder and cobble conglomerates exposed here are correlative with the Maestrichtian part of the Evanston Formation seen on Day 9. These strata onlay the Jurassic rocks with an angular unconformity and show provenance directly from the uplift. Combined with the Echo Canyon Conglomerate, the Devils Slide alluvial fan deposits demonstrate a sequential uplift history for the basement ramp anticline that is contemporaneous with movement on the major thrusts to the east. The Maestrichtian conglomerates are readily accessible along the Weber River Canyon at Devils Slide.

Stop 40

West from Devils Slide, the basement uplift is cut by a Basin and Range normal fault at Morgan (Figure 97). In

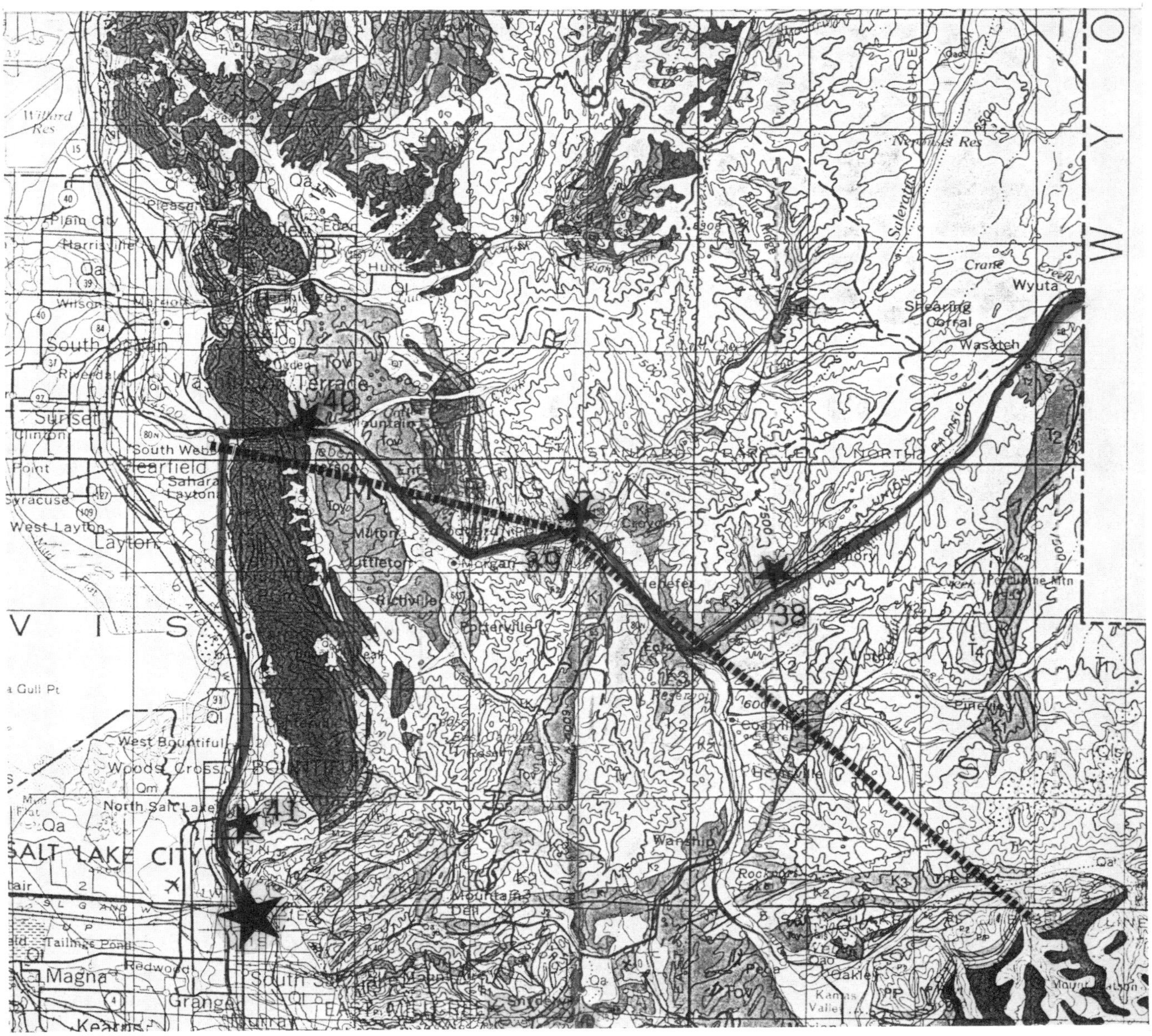

FIGURE 97 Segment of the geologic map of Utah (Hintze, 1980), showing the last part of the route on Day 10 (heavy black line). The large star marks the end of Trip T-328 at Salt Lake City. The smaller numbered stars mark stops along the route. The dashed line is the line of cross section shown in Figure 98. For geologic symbols and line patterns see source map. Squares are approximately six miles on a side.

the Morgan Valley, the Oligocene volcaniclastic sequence in the graben is exposed with 15° to 50° east dips into the normal fault. The west side of the valley is bounded by Archean crystalline rocks of the basement uplift, which are well exposed along the Weber River. The crystalline basement is cut by numerous ductile and brittle shear zones that record internal deformation both prior to and during major uplift of the basement (Yonkee and Bruhn, 1987; Yonkee, and others, 1989).

The route turns sharply to the south just west of the canyon mouth and in about 10 miles joins U.S. Highway 89 to Salt Lake City. This is the final segment of the field trip and it exhibits an entirely different type of stratigraphy, sedimentation, and tectonism than was observed on the previous nine days. The conspicuous faceted mountain front and numerous Holocene fault scarps are the main surface expressions of a system of normal fault segments that have locally accommodated at least seven miles of vertical displacement (Parry and Bruhn, 1987). The topographic terraces along the mountain front are expressions of ancient shoreline levels of Pleistocene Lake Bonneville. The Pleistocene deltaic sequences so dramatically developed here are at the site of the pioneering classic descriptions of deltaic sedimentation by Gilbert

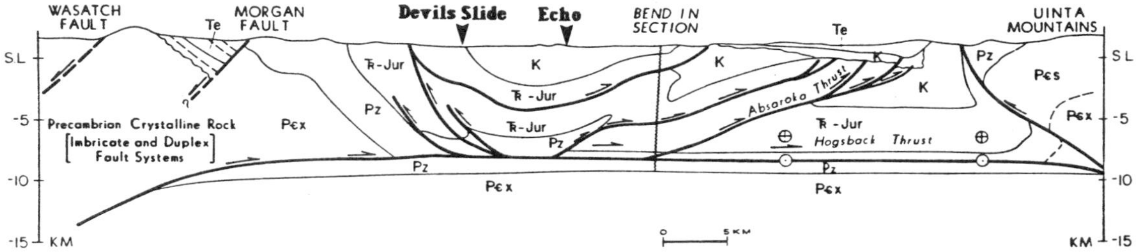

FIGURE 98 Cross section between the north flank of the Uinta Mountains and the Wasatch fault. The locations of the Echo Stop 38 and the Devils Slide Stop 39 are labeled. Note the relationship between the Wasatch basement uplift and decollement thrusts to the east. See Figure 97 for location of section. From Bruhn, and others, (1986) after Royse and Lamerson (1984).

(1890). This is also an area of ongoing seismic activity (Smith and Sbar, 1974, and many subsequent reports). In addition, because of the ever-increasing population density along the Wasatch Mountain front, many land use planning studies related to the faults have been made. This is not the place to present a literature survey but an example of one such study is a two volume report by Cluff, and others, (1974).

Salt Lake City

This is the end of trip T-328. The compilers of this report hope you have shared with them, in the field, the excitement and appreciation of the geology of this region as it has happened and is still happening. There is much yet to be learned here so please apply your knowledge of other areas to the mysteries that remain in this one. Bon Voyage!

Acknowledgments

This narrative could not have been done without the superb cooperation of the local experts listed at the beginning of this report. In addition, in alphabetic order, the following talented and enthusiastic people have contributed immeasurably in this production: Dr. D.L. Blackstone, Jr., Geology Department, University of Wyoming; Dr. William G. Brown, professor and field geologist, Baylor University; Betty Doerges, skillful, patient and ingenious compositor; Carol Edwards, Field Records, U.S.G.S.; Dr. Jack Fanshawe, the "Grand Old Man" of Bighorn-Beartooth-Owl Creek geology; Dr. George Frison, Wyoming Archeologist; Richard W. Jones, coal expert, Geological Survey of Wyoming; Bob Lynn, photographer; Dr. Malcolm C. McKenna, for providing key photographs; Robert Peck, publisher, Riverton Ranger; Jack Rathbone, geologist-photographer; Dr. Marith Reheis, geomorphologist, U.S.G.S.; Juergen Reinhardt, Penelope Hanshawe, and John C. Reed, Jr., U.S.G.S., for the I.G.C. program skills; Phyllis Ranz, Geological Survey of Wyoming, for the beautiful and appropriate cover illustration; Jack Richard, outstanding photographer of mountains and basins; Sheila Roberts, Geological Survey of Wyoming, for much technical advice on the manuscript; Dr. John S. Shelton, geologist-photographer; and Ned Sterne, AMOCO geologist. Last, but far from least, my gratitude goes to geologist Jane M. Love for her infinite patience, skill as a critic and editor, and for being a true partner in this enterprise.

REFERENCES CITED

Albee, H.F., 1968, Geologic map of the Munger Mountain Quadrangle, Teton and Lincoln Counties, Wyoming: U.S. Geol. Survey Geologic Quadrangle Map GQ-705.

Bargar, K.E., 1978, Geology and thermal history of Mammoth Hot Springs, Yellowstone National Park, Wyoming: U.S. Geol. Survey Bull. 1444, 55 p.

Barlow, J.A., and Haun, J.D., 1978, Oil and gas production, reserves, and resouces in Wyoming: Minerals Division, Wyo. Dept. of Economic Planning and Development, 18 p.

Barnosky, A.D., 1984, The Colter Formation: evidence for Miocene volcanism in Jackson Hole, Teton County, Wyoming: Earth Science Bull., v. 17, 1984, p. 49-97.

Blackstone, D.L., Jr., 1986, Structural geology – northwest margin, Bighorn Basin: Park County, Wyoming and Carbon County, Montana: Montana Geol. Soc. and Yellowstone. Bighorn Research Assoc. joint field conference and symposium: p. 125-135.

______, 1988, Traveler's guide to the geology of Wyoming: Geol. Survey of Wyo. Bull. 67, 130 p.

______1989, Precambrian basement map of Wyoming: outcrop and structural configuration: Geol. Survey of Wyo. map series 27.

Blackstone, D.L., Jr., and De Bruin, R.H., 1987, Tectonic map of the overthrust belt, western Wyoming, northwestern Utah, and southeastern Idaho: Wyoming Geologic Survey Map MS 23.

Bonini, W.E., and Kinard, R.E., 1983, Gravity anomalies along the Beartooth front, Montana: evidence for a low-angle thrust: Wyo. Geol. Assoc. Guidebook 34th Ann. Field Conf.: p. 89-94.

Bonini, W.E., Fox S.K., and Judson, Sheldon, 1986, The Red Lodge project and the YBRA: the early years, 1930-1942: Montana Geol. Soc. and Yellowstone Bighorn Research Assoc. joint field conference and symposium: p. 1-9.

Bradley, W.H., 1964, Geology of Green River Formation and associated Eocene rocks in southwestern Wyoming and adjacent parts of Colorado and Utah: U.S. Geol. Survey Prof. Paper 496-A, p. A1-A86.

Breckenridge, R.M., and Hinckley, B.S., 1978, The thermal springs of Wyoming: Geol. Survey of Wyoming Bull. 60, 104 p.

Brown, C.W., 1961, Cenozoic stratigraphy and structural geology, northeastern Yellowstone National Park, Wyoming and Montana: Geol. Soc. America Bull., v. 72, no. 8, p. 1173-1193.

Brown, W.G., 1984, A reverse fault interpretation of Rattlesnake Mountain anticline, Bighorn Basin, Wyoming: Mountain Geologist, v. 21, p. 31-37.

Bruhn, R.L., Picard, M.D., and Isby, J.S., 1986, Tectonics and sedimentology of Uinta Arch, western Uinta Mountains, and Uinta basin: *in* Peterson, J.A. (ed.), Paleotectonics and sedimentation: Am. Assoc. Petrol. Geol. Memoir 41, p. 333-352.

Bucher, W.H., 1947, Heart Mountain problem, *in* Wyo. Geol. Assoc. Guidebook 2nd Ann., Field Conf.,: p. 189-197.

Christiansen, R.L., 1984, Yellowstone magmatic evolution: its bearing on understanding large-volume explosive volcansim, *in* Explosive volcanism: inception, evolution, hazards; Studies in Geophysics: National Academy Press, Washington, D.C., p. 84-95.

Christiansen, R.L. and Blank, H.R., Jr., 1972, Volcanic stratigraphy of the Quaternary rhyolite plateau in Yellowstone National Park: U.S. Geol. Survey Prof. Paper 729-B, p. B1-B18.

______, 1974, Geologic map of the Madison Junction quadrangle, Yellowstone National Park, Wyoming: U.S. Geol. Survey Geologic Quadrangle Map GQ-1190.

______, 1975, Geologic map of the Canyon Village quadrangle, Yellowstone National Park, Wyoming: U.S. Geol. Survey Geologic Quadrangle Map GQ-1192.

Cluff, L.S., Glass, C.E., and Brogan, G.E., 1974, Investigation and evaluation of the northern Wasatch and Cache Valley faults – a guide to land-use planning with recommendations for seismic safety; prepared for U.S. Geol. Survey by Woodward-Lundgren and Associates, 147 pages.

Comstock, T.B., 1875, Geological report, *in* Jones, W.A., Report upon the reconnaissance of northwestern Wyoming, including Yellowstone National Park, made in the summer of 1873: Washington, U.S. Govt. Printing Office, p. 85-292.

Coogan, J.C., 1987, Thrust systematics and displacement transfer in the Wyoming-Idaho-Utah thrust belt: Geol. Soc. of America (Abs.), v. 19, no. 7, p. 626.

Coogan, J.C., and Boyer, S.E., 1985, Variable shortening mechanisms in thrust belts and implications for balanced cross section construction; an example from the Idaho-Wyoming thrust belt: Geol. Soc. America (Abs.), v. 17, no. 7, p. 552.

Coogan, J.C., and Yonkee, W.A., 1985, Salt detachments within the Meade and Crawford thrust systems, Idaho and Wyoming, *in* Kerns, G.J., and Kerns, R.L., eds., Orogenic patterns and stratigraphy of north-central Utah and southeastern Idaho: Utah Geol. Assoc. Pub. 14, p. 75-82.

DeCelles, P.G., 1988, Lithologic provenance modeling applied to the Late Cretaceous synorogenic Echo Canyon Conglomerate, Utah: A case of multiple source areas: Geology, v. 16, p. 1039-1043.

Degenkolb, H.J., 1971, Preliminary structural lessons from the earthquake, *in* The San Fernando, California, earthquake of February 9, 1971: U.S. Geol. Survey Prof. Paper 733, p. 133-134.

Dorf, E., 1964, The petrified forests of Yellowstone National Park. Washington, D.C., Govt. Printing Office, 12 p.

Dorr, J.A., Jr., Spearing, D.R., and Steidtmann, J.R., 1977, Deformation and deposition between a Foreland Uplift and an impinging Thrust Belt: Hoback Basin, Wyoming. Geol Soc. America Sp. Paper 177, 82 p.

Dunn, S.L.D., 1983, Timing of foreland and thrust-belt deformation in a overlap area, Teton Pass, Idaho and Wyoming: Rocky Mountain Assoc. Geologists; Rocky Mountain foreland basins and uplifts, J.D. Lowell, Ed.: p. 263-269.

Dutcher, L.A.F., Jobling, J.L., and Dutcher, R.R., 1986, Stratigraphy, sedimentology, and structural geology of Laramide synorogenic sediments marginal to the Beartooth Mountains, Montana and Wyoming: Montana Geol. Soc. and Yellowstone Bighorn Research Assoc. joint field conference and symposium: p. 33-52.

Fanshawe, J.R., 1939, Structural geology of Wind River Canyon area, Wyoming: Am. Assoc. Petrol. Geol. Bull., v. 23, no. 10, p. 1439-1492.

Foose, R.M., Wise, D.U., and Garbarini, G.S., 1961, Structural geology of the Beartooth Mountains, Montana, and Wyoming: Geol. Soc. America Bull., v. 72, p. 1143-1172.

Foose, R.M., Wise, D.U., Fanshawe, J.R., Dutcher, L.A.F., Dutcher, R.R., and French, D.E., 1986, Road log: Red Lodge to Dead Indian Hill via Clarks Fork Canyon: tectonics and geologic history of the Beartooth Uplift: Montana Geol. Soc. and Yellowstone Bighorn Research Assoc. joint field conference and symposium: p. 291-306.

Fraser, G.D., Waldrop, H.A., and Hyden, H.J., 1969, Geology of the Gardiner area, Montana: U.S. Geol. Survey Bull. 1277, 118 p.

Frison, G.C., 1978, Prehistoric hunters of the High Plains: Academic Press, 457 p.

Fritz, W.J., 1982, Geology of the Lamar River Formation, northeastern Yellowstone National Park: Wyo. Geol. Assoc. 33rd Ann. Field Conf.: p. 73-101.

Gilbert, G.K., 1890, Lake Bonneville: U.S. Geol. Survey Monograph 1, 438 p.

Glass, G.B., and Roberts, J.T., 1978, Update on the Wind River coal basin: Wyo. Geol. Assoc. Guidebook 30th Ann. Field Conf., p. 363-377.

Grande, Lance, 1980, Paleontology of the Green River Formation, with a review of the fish fauna: Geol. Survey of Wyoming Bull. 63, 333 p.

Hauge, T.A., 1985, Gravity-spreading origin of the Heart Mountain allochthon, northwestern Wyoming: Geol. Soc. America Bull., v. 96, p. 1440-1456.

Hayden, F.V., 1872, Preliminary report of the U.S. Geological Survey of Wyoming and portions of contiguous territories (being a second annual report of progress for 1870): Washington, U.S. Govt. Printing Office, 511 p.

Hinckley, B.S., Heasler, H.P., and King, J.K., 1982, The Thermopolis hydrothermal system with an analysis of Hot Springs State Park: Geol. Survey of Wyoming Prelim. Rept. No. 20, 42 p.

Hintze, L.F., 1980, Geological map of Utah: Utah Geological and Mineral Survey.

Hsu, K.J., 1969, Role of cohesive strength in the mechanics of overthrust faulting and of landsliding: Geol. Soc. America Bull., v. 80, p. 927-952.

Hubbert, M.K., and Rubey, W.W., 1959, Role of fluid pressure in mechanics of overthrust faulting, I. Mechanics of fluid-filled porous solids and its application to overthrust faulting: Geol. Soc. America Bull., v. 70, p. 115-166.

Hughes, C.J., 1970, The Heart Mountain detachment fault – A volcanic phenomenon?: Journal of Geology, v. 78, p. 107-116.

Hunter, R.B., 1987, Timing and structural relations between the Gros Ventre foreland uplift, the Prospect thrust system, and the Granite Creek thrust, *in* Miller, W.R. (ed.), The Thrust belt revisited: Wyo. Geol. Assoc. Guidebook 38th Ann. Field Conf., p. 109-131.

Hurst, D.J., and Steidtmann, J.R., 1986, Stratigraphy and tectonic significance of the Tunp conglomerate in the Fossil Basin, southwest Wyoming: Mountain Geologist, v. 23, p. 6-13.

Keefer, W.R., 1972, The geologic story of Yellowstone National Park: U.S. Geol. Survey Bull. 1347, 92 p.

Kehle, R.O., 1970, Analysis of gravity sliding and orogenic translation: Geol. Soc. America Bull., v. 81, p. 1641-1664.

Knittel, Peggy, 1974, A field guide to the Alcova area: Wyo. Field Science Foundation, 60 p.

Kvenvolden, K.A., Simoneit, B.R.T., and Love, J.D., Natural gas seeps in Yellowstone and Grand Teton National Parks, Wyoming: Wyo. Geol. Assoc. Guidebook 40th Ann. Field Conf.: (in press).

Lageson, D.R., 1980, Structural geology of the Stewart Peak quadrangle, Lincoln County, Wyoming, and adjacent parts of the Idaho-Wyoming thrust belt, unpublished Ph.D. dissertation, University of Wyoming, 293 p.

______, 1984, Structural geology of the Stewart Peak culmination, Idaho-Wyoming thrust belt. Am. Assoc. Petrol. Geol. Bull., v. 15, p. 401-406.

Lageson, D.R., and Spearing, D.R., 1988, Roadside geology of Wyoming: Mountain Press Publishing Co., 271 p.

Lamerson, P.R., 1982, The Fossil Basin and its relationship to the Absaroka thrust system, Wyoming and Utah, *in* R.B. Powers (ed.), Geologic studies of the Cordilleran thrust belt. Rocky Mtn. Assoc. Geol, p. 279-340.

Loose, S.A., and Boberg, W.W., 1987, Sedimentary facies control on mineralization at the Lake Alice district in the Wyoming overthrust belt: Wyo. Geol. Assoc. Guidebook 38th Ann. Field Conf.: p. 309-327.

Love, J.D., 1939, Geology along the southern margin of the Absaroka Range, Wyoming: Geol. Soc. America Sp. Paper 20, 134 p.

______,1954, Preliminary report on uranium in the Gas Hills area, Fremont and Natrona Counties, Wyoming: U.S. Geol. Survey Circular 352, 11 p.

______, 1961, Reconnaissance study of Quaternary faults in and south of Yellowstone National Park, Wyoming: Geol. Soc. America Bull., v. 72, no. 12, p. 1749-1764.

______, 1970, Cenozoic geology of the Granite Mountains area, central Wyoming: U.S. Geological Survey Prof. Paper 495-C, 154 p.

______, 1977, Summary of Upper Cretaceous and Cenozoic stratigraphy and of tectonic and glacial events in Jackson Hole, northwestern Wyoming, *in* Rocky Mountain thrust belt geology and resources, Wyo. Geol. Assoc. Guidebook 29th Ann. Field Conf.: p. 585-593.

______, 1984, Gold, silver, and other selected trace elements in the Phosphoria Formation of western Wyoming. Wyo. Geol. Assoc. Guidebook 35th Ann. Field Conf., 1984, p. 383-387.

______, 1987, Teton Mountain front, northwestern Wyoming: Geological Society of America Centennial Field Guide Volume 2 (DNAG), Rocky Mountain Section, p. 173-178.

______, 1988a, Geology of the Wind River Basin, central Wyoming: DNAG volume Sedimentary Cover, North American Craton: p. 196-200.

______, 1988b, Geology of the Bighorn Basin, northern Wyoming and southeastern Montana: DNAG volume Sedimentary Cover, North American Craton. p. 201-204.

Love, J.D. and Albee, H.F., 1972, Geologic map of the Jackson quadrangle, Teton County, Wyoming: U.S. Geol. Survey Miscel. Geologic Invest. Map I-769-A.

Love, J.D. and Antweiler, J.C., 1973, Copper, silver, and zinc in the Nugget Sandstone, western Wyoming: Wyo. Geol. Assoc. Guidebook 25th Ann. Field Conf., 1973: p. 139-147.

Love, J.D. and Christiansen, A.C., 1980, Preliminary correlation of stratigraphic units used on 1° x 2° geologic quadrangle maps of Wyoming: Wyo. Geol. Assoc. Guidebook 31st Ann. Field Conf., 1980, p. 279-282.

______, 1985, Geologic map of Wyoming. U.S. Geological Survey Map.

Love, J.D., and Keefer, W.R., 1969, Basin Creek uplift and Heart Lake Conglomerate, southern Yellowstone National Park, Wyoming, *in* Geological Survey Research 1969: U.S. Geol. Survey Prof. Paper 650-D, p. D122-D130.

______, 1975, Geology of sedimentary rocks in southern Yellowstone National Park, Wyoming: U.S. Geol. Survey Prof. Paper 729-D, p. D1-D60.

Love, J.D., Leopold, E.B., and Love, D.W., 1978, Eocene rocks, fossils, and geologic history, Teton Range, northwestern Wyoming, *in* Geology of the Teton-Jackson Hole region, northwestern Wyoming: U.S. Geol. Survey Prof. Paper 932-B, 40 p.

Love, J.D., and Love, J.M., 1983, Road log, Jackson to Dinwoody and return: Geological Survey of Wyoming Public Information Circular No. 20, 34 p.

______, 1988, Road log, north end of Blacktail Butte up Gros Ventre River Valley to Crystal Creek and return along Kelly road: Geological Survey of Wyoming Reprint 46, 14 p.

Love, J.D. and Montagne, J.M. de la, 1956, Pleistocene and recent tilting of Jackson Hole, Teton County, Wyoming. Wyo. Geol. Assoc. Guidebook 11th Ann. Field Conf.: p. 169-178.

Love, J.D., and Reed, J.C., Jr., 1971, Creation of the Teton landscape – the geologic story of Grand Teton National Park: Grand Teton Natural History Association, 120 p.

Love, J.D., Reed, J.C., Jr., Christiansen, R.L., and Stacy, J.R., 1973, Geologic block diagram and structural history of the Teton region, northwestern Wyoming: U.S. Geological Survey Miscel. Geologic Invest. Map I-730.

Love, J.D., Reed, J.C. Jr., and Christiansen, A.C., Geologic map of Grand Teton National Park: U.S. Geological Survey (in press).

Love, L.L., Kudo, A.M., and Love, D.W., 1976, Dacites of Bunsen Peak, the Birch Hills, and the Washakie Needles, northwestern Wyoming, and their relationship to the Absaroka volcanic field, Wyoming and Montana: Geol. Soc. America Bull., v. 87, p. 1455-1462.

Lovering, T.S., 1929, The New World or Cooke City mining district, Park County, Montana: U.S. Geol. Survey Bull. 811-A, p. 1-87.

MacGinitie, H.D., 1969, The Eocene Green River flora of northwestern Colorado and northeastern Utah: California Univ. Pubs. Geol. Sci., v. 83, 203 p.

McKenna, M.C., and Love, J.D., 1970, Local stratigraphic and tectonic significance of *Leptoceratops,* a Cretaceous dinosaur in the Pinyon Conglomerate, northwestern Wyoming, *in* U.S. Geological Survey Research 1970: U.S. Geol. Survey Prof. Paper 700-D, p. D55-D61.

______, 1972, High-level strata containing early Miocene mammals on the Bighorn Mountains, Wyoming: American Museum Novitates no. 2490, 31 p.

Maley, R.P., and Cloud, W.K., 1971, Preliminary strong-motion results from the San Fernando earthquake of February 9, 1971, *in* The San Fernando, California, earthquake of February 9, 1971: U.S. Geol. Survey Prof. Paper 733, p. 163-176.

Mansfield, G.R., 1927, Geography, geology, and mineral resources of a part of southeastern Idaho: U.S. Geol. Survey Prof. Paper 152, 453 p.

Marler, G.D., 1964, Effects of the Hebgen Lake earthquake of August 17, 1959, on the hot springs of the Firehole Geyser Basins, Yellowstone National Park: U.S. Geol. Survey Prof. Paper 435, p. 185-197.

______, 1969, The story of Old Faithful: Yellowstone Library and Museum Association: 49 p.

Meyer, G.A., 1986, Genesis and deformation of Holocene shoreline terraces, Yellowstone Lake, Wyoming: Montana State Univ. M.S. thesis (unpub.), 94 p.

Meyer, J.E., 1984, The stratigraphy of the Gypsum Spring Formation (Middle Jurassic) northwestern Bighorn Basin, Wyoming: Univ. Wyoming M.S. thesis (unpub.), 93 p.

Mills, J.P., 1989, Foreland structure and karstic ground water circulation in the eastern Gros Ventre Range, Wyoming. Univ. of Wyo. M.S. thesis (unpub.), 101 p.

Morrill, B.J., 1971, Evidence of record vertical accelerations at Kagel Canyon during the earthquake *in* The San Fernando, California, earthquake of February 9, 1971: U.S. Geol. Survey Prof. Paper 733, p. 177-181.

Muffler, L.J.P., White, D.E., and Truesdell, A.H., 1971, Hydrothermal explosion craters in Yellowstone National Park: Geol. Soc. America Bull., v. 82, p. 723-740.

Nason, R.D., 1971, Shattered earth at Wallaby Street, Sylmar, *in* The San Fernando, California, earthquake of February 9, 1971: U.S. Geol. Survey Prof. Paper 733, p. 97-98.

National Research Council, Committee on Earthquake Engineering Research, 1969, Earthquake Engineering Research: Natl. Acad. Sci., p. 5 (available as PB 188 636 from Natl. Tech. Inf. Service, U.S. Dept. Commerce, Springfield, VA, 22151).

Ori, G.G., and Friend, P.F., 1984, Sedimentary basins formed and carried piggyback on active thrust sheets: Geology, v. 12, p. 475-478.

Oriel, S.S., Antweiler, J.C., and Moore, D.W., 1985, Mineral resource potential map of the west and east Palisades Roadless Areas, Idaho and Wyoming: U.S. Geol. Survey Map MF-1691.

Oriel, S.S. and Tracey, J.I., 1970, Uppermost Cretaceous and Tertiary Stratigraphy of Fossil Basin, southwestern Wyoming: U.S. Geol. Survey Prof. Paper 635, 53 p.

Osterwald, F.W., 1961, Critical review of some tectonic problems in Cordilleran foreland: Am. Assoc. Petrol. Geol. Bull., v. 45, no. 2, p. 219-237.

Ostrom, J.H., 1970, Stratigraphy and paleontology of the Cloverly Formation (Lower Cretaceous) of the Bighorn Basin area, Wyoming and Montana: Peabody Museum of Natural History, Yale University Bull. 35, 234 p.

Palmquist, Robert, 1983, Terrace chronologies in the Bighorn Basin, Wyoming: Wyo. Geol. Assoc. Guidebook 34th Ann. Field Conf.. p. 217-231.

Parry, T.W., and Bruhn, R.L., 1987, Fluid inclusion evidence for minimum 11 km vertical offset on the Wasatch fault, Utah: Geology, v. 15, p. 67-70.

Pierce, K.L., 1979, History and dynamics of glaciation in the northern Yellowstone National Park area: U.S. Geol. Survey Prof. Paper 729-F, p. F1-F90.

Pierce, W.G., 1963, Reef Creek detachment fault, northwestern Wyoming: Geol. Soc. America Bull., v. 74, p. 1225-1236.

______, 1966, Geologic map of the Cody Quadrangle, Park County, Wyoming: U.S. Geol. Survey Geologic Quadrangle Map GQ-542.

______, 1973, Principal features of the Heart Mountain fault and the mechanism problem, *in* DeJong, K.A., and Scholten, Robert, eds., Gravity and tectonics: New York, John Wiley & Sons, p. 457-471.

______, 1979, Clastic dikes of Heart Mountain fault breccia, northwestern Wyoming, and their significance: U.S. Geol. Survey Prof. Paper 1133, 25 p.

______, 1986, The South Fork detachment fault, Park County, Wyoming: discussion and reply: Contrib. to Geology, University of Wyoming, v. 24, no. 1, p. 77-90.

______, 1987a, Detachment, a faulting process (Décollement) *in* Seyfert, C., ed., The Encyclopedia of Structural Geology and Plate Tectonics: New York, Van Nostrand Reinhold, p. 192-198.

______, 1987b, The case for tectonic denudation by the Heart Mountain fault – A response: Geol. Soc. America Bull., v 99, p. 552-568.

Pierce, W.G., and Nelson, W.H., 1971, Geologic map of the Beartooth Butte Quadrangle, Park County, Wyoming: U.S. Geol. Survey Geologic Quadrangle Map GQ-935, scale 1:62,500.

______, 1973, Crandall Conglomerate, an unusual stream deposit, and its relation to Heart Mountain faulting: Geol. Soc. America Bull., v. 84, p. 2631-2644.

Pierce, W.G., Nelson, W.H., and Prostka, H.J., 1973, Geologic map of the Pilot Peak quadrangle, Park County, Wyoming: U.S. Geol. Survey Miscel. Geologic Invest. Map I-816, scale 1:62,500.

Reed, J.C., Jr., and Zartman, R.E., 1973, Geochronology of Precambrian rocks of the Teton Range, Wyoming: Geol. Soc. America Bull., v. 84, no. 2, p. 561-582.

Reheis, M.C., 1987, Soils in granitic alluvium in humid and semiarid climates along Rock Creek, Carbon County, Montana: U.S. Geol. Survey Bull. 1590-D, p. D1-D71.

Richards, P.W., 1955, Geology of the Bighorn Canyon-Hardin area, Montana and Wyoming: U.S. Geol. Survey Bull. 1026, 93 p.

Richmond, G.M., 1976, Surficial geologic history of the Canyon Village Quadrangle, Yellowstone National Park, Wyoming, for use with Map I-652. U.S. Geol. Survey Bull. 1427, 35 p.

______, 1987, Geology and evolution of the Grand Canyon of the Yellowstone, Yellowstone National Park, Wyoming: Geol. Soc. America Centennial Field Guide – Rocky Mountain Section, p. 155-160.

Rosholt, J.N., 1976, $^{230}Th/^{234}U$ dating of travertine and caliche rinds: Geol. Soc. America Abs. with Programs, v. 8, no. 6, p. 1076.

Royse, F., Jr., and Lamerson, P.R., 1984, Field trip guide, northern Utah thrust belt and north Cottonwood uplift: Am. Assoc. Petrol. Geol. Structural Geology School, Park City, Utah.

Royse, F., Jr., Warner, M.A., and Reese, D.L., 1975, Thrust belt structural geometry and related stratigraphic problems, Wyoming-Idaho-northern Utah, *in* D.W. Bolyard (ed.), Deep drilling frontiers of the central Rocky Mountains. Rocky Mtn. Assoc. Geol., p. 41-54.

Rubey, W.W., and Hubbert, M.K., 1959, Role of fluid pressure in mechanics of overthrust faulting, II. Overthrust belt in geosynclinal area of western Wyoming in light of fluid-pressure hypothesis. Geol. Soc. America Bull., v. 70, p. 167-205.

Ruppel, E.T., 1972, Geology of Pre-Tertiary rocks in the northern part of Yellowstone National Park, Wyoming: U.S. Geol. Survey Prof. Paper 729-A, p. A1-A66.

Salat, T.S., 1989, Provenance, dispersal, and tectonic significance of the Evanston Formation and Sublette Range Conglomerate, Idaho-Wyoming-Utah: Univ. Wyoming M.S. thesis (unpub.).

Schmitt, J.G., 1987, Origin of late Cretaceous to early Tertiary quartzite conglomerates – northwestern Wyoming: Thirty-eighth Field Conference, *in* Miller, W.R. (ed.), The thrust belt revisited: Wyo. Geol. Assoc. Guidebook 38th Ann. Field Conf., p. 217-224.

Schroeder, M.L., 1969, Geologic map of the Teton Pass Quadrangle, Teton County, Wyoming: U.S. Geol. Survey Geologic Quadrangle Map GQ-793.

______, 1972, Geologic map of the Rendezvous Peak Quadrangle, Teton County, Wyoming: U.S. Geol. Survey Geologic Quadrangle Map GQ-980.

______, 1976, Geologic Map of Bull Creek Quadrangle, Teton and Sublette Counties, Wyoming: U.S. Geol. Survey Geologic Quadrangle Map GQ-1300.

Schultz, A.R., 1914, Geology and geography of a portion of Lincoln County, Wyoming: U.S. Geol. Survey Bull. 543, 141 p.

______, 1918, A geologic reconnaissance for phosphate and coal in southeastern Idaho and western Wyoming: U.S. Geol. Survey Bull. 680, 84 p.

Smedes, H.W., and Prostka, H.J., 1972, Stratigraphic framework of the Absaroka Volcanic Supergroup in the Yellowstone National Park region: U.S. Geol. Survey Prof. Paper 729-C, p. C1-C33.

Smith, R.B., and Braile, L.W., 1984, Crustal structure and evolution of an explosive silicic volcanic system at Yellowstone National Park, *in* Explosive volcanism: inception, evolution, and hazards; Studies in Geophysics: National Academy Press, Washington, D.C., p. 96-109; (reprinted from Wyo. Geol. Assoc. Guidebook 33rd Ann. Field Conf., 1982, p. 233-250).

Smith, R.B., and Sbar, M.L., Contemporary tectonics and seismicity of the western United States with emphasis on the Intermountain Seismic Belt: Geol. Soc. America Bull., v. 85, p. 1205-1218.

Tourtelot, H.A., and Thompson, R.M., 1948, Geology of the Boysen area, central Wyoming: U.S. Geol. Survey Oil and Gas Inv. Prelim. Map 91.

Tracey, J.I., Jr., Oriel, S.S., and Rubey, W.W., 1961, Diamictite facies of the Wasatch Formation in the Fossil Basin, southwestern Wyoming, *in* Short papers in the geologic and hydrologic sciences: U.S. Geol. Survey Prof. Paper 424-B, p. B149-B150.

University of Wyoming, 1971, Trona Issue: Contrib. to Geology, v. 10, no. 1, 72 p.

U.S. Geological Survey, 1964, The Hebgen Lake, Montana earthquake of August 17, 1959: U.S. Geol. Survey Prof. Paper 435, 242 p.

U.S. Geological Survey, 1972, Geologic map of Yellowstone National Park: U.S. Geol. Survey Map.

Valenti, G.L., 1982, Structure of the Laketown quadrangle and petroleum exploration in the Bear Lake area, Rich County, Utah, *in* R.B. Powers (ed.), Geologic studies of the Cordilleran thrust belt: Rocky Mtn. Assoc. Geol, p. 859-868.

Van Burgh, D.P., Jr., 1978, A field guide to the Casper Mountain area: Wyo. Field Science Foundation, 80 p.

Veatch, A.C., 1907, Geography and geology of a portion of southwestern Wyoming, with special reference to oil and coal: U.S. Geol. Survey Prof. Paper 56, 128 p.

Voight, Barry, 1978, Lower Gros Ventre Slide, Wyoming, U.S.A., *in* Rockslides and avalanches I, Natural phenomena: Elsevier Publishing Company, New York, p. 113-166.

Warner, M.A., 1982, Source and time of generation of hydrocarbons in the Fossil Basin, western Wyoming thrust belt, *in* R.B. Powers (ed.), Geologic studies of the Cordilleran thrust belt: Rocky Mtn. Assoc. Geol., p. 805-815.

Waterman, H.D., 1983, Road log, U.S. Highway 14-A to Devils Canyon overlook in the Bighorn Canyon recreational area: Wyo. Geol. Assoc. Guidebook 34th Ann. Field Conf: p. 266-269.

White, D.E, Hutchinson, R.A., and Keith, T.E.C., 1988, The geology and remarkable thermal activity of Norris Geyser Basin, Yellowstone National Park, Wyoming: U.S. Geol. Survey Prof. Paper 1456, 84 p.

Wiltschko, D.V., and Dorr, J.A., 1983, Timing of deformation in overthrust belt and foreland of Idaho, Wyoming, and Utah: Am. Assoc. Petrol. Geol. Bull., 67, p. 1304-1322.

Wise, D.U., 1983, Overprinting of Laramide structural grains in the Clarks Fork Canyon area and eastern Beartooth Mountains of Wyoming: Wyo. Geol. Assoc. Guidebook 34th Ann. Field Conf.: p. 77-87.

Witkind, I.J., 1964, Events on the night of August 17, 1959 – the human story, *in* The Hebgen Lake, Montana Earthquake of August 17, 1959: U.S. Geol. Survey Prof. Paper 435, p. 1-4.

Wold, R.W., Mayhew, M.A., and Smith, R.B., 1977, Bathymetric and geophysical evidence for a hydrothermal explosion crater in Mary Bay, Yellowstone Lake, Wyoming: Jour. Geoph. Research, v. 82, no. 26, p. 3733-3738.

Wyoming Geological Association, 1983, Geology of the Bighorn Basin: Wyo. Geol. Assoc. Guidebook 34th Ann. Field Conf.: 274 p.

Yonkee, W.A., and Bruhn, R.L., 1987, Basement-cover relations, Sevier orogenic belt, northern Utah: Geol. Soc. America (Abs.), v. 19, p. 902.

Yonkee, W.A., Parry, W.T., Bruhn, R.L., and Cashman, P.H., 1989, Thermal models of thrust faulting: Constraints from fluid-inclusion observations, Willard thrust sheet, Idaho-Utah-Wyoming thrust belt: Geol. Soc. America Bull., v 101, p. 304-315.

Zapan, Shirley, and Owens, H.J., 1979, Red Lodge, saga of a western area: Carbon County Historical Society, Frontier Press, Billings, Montana, 410 p.

Zeller, C.G., 1982, Structural geology along Teton Pass, Wyoming, *in* Geologic studies of the Cordilleran thrust belt, R.B. Powers, ed.: Rocky Mtn. Assoc. Geol., v. 2, p. 831-843.